ISBN 978-1-330-34459-0
PIBN 10034789

1 MONTH OF
FREE
READING

at
www.ForgottenBooks.com

By purchasing this book you are eligible for one month membership to ForgottenBooks.com, giving you unlimited access to our entire collection of over 1,000,000 titles via our web site and mobile apps.

To claim your free month visit:
www.forgottenbooks.com/free34789

COLOUR IN NATURE

A STUDY IN BIOLOGY

BY

MARION I. NEWBIGIN

D.Sc. (LOND.)

LECTURER ON ZOOLOGY
IN THE MEDICAL COLLEGE FOR WOMEN, EDINBURGH

LONDON

JOHN MURRAY, ALBEMARLE STREET

PREFACE

THIS little book is an attempt to set forth in systematic order the main facts at present known in regard to the Pigments and Colours of Plants and Animals. Although there are not a few books which treat of the external aspects of the Colours of organisms, it is a frequent complaint among those engaged in actual research that, not only is the physiological side of the subject inadequately treated in the text-books, but the literature is also scattered and difficult of access. It is hoped that the present work may both help to supply this deficiency, and also appeal to that wider public whose interest in the problems of Colour has been aroused by the existing general books on the subject. The popularity of many of these is deservedly so wide, that I have endeavoured throughout to avoid trenching on ground already covered by them, and have rather concentrated attention upon less familiar points.

13/4 ᵉᵉ

The literature of the subject is so extensive that it has been found impossible within the assigned limits of space to attempt a complete bibliography. I have therefore thought it most advisable to confine myself as far as possible in the list of references to papers actually consulted and discussed in the text. Even thus abbreviated, the list will, I trust, be of value to the workers in the subject.

I have to express my sincere obligations to Mr. J. A. Thomson for much kind assistance, especially with the literature.

COLLEGE OF MEDICINE FOR WOMEN,
EDINBURGH, *September* 1898.

CONTENTS

INTRODUCTION

CHAPTER I

THE COLOURS OF ORGANISMS

CHAPTER II

THE PIGMENTS OF ORGANISMS

CHAPTER VI

THE COLOURS OF CRUSTACEA AND ECHINODERMA

CHAPTER VII

THE COLOURS OF THE LEPIDOPTERA

CHAPTER VIII

THE COLOURS OF INSECTS IN GENERAL AND OF SPIDERS

CHAPTER IX

THE COLORATION OF MOLLUSCA AND OF INVERTEBRATES IN GENERAL

CHAPTER X

THE COLOURS OF FISHES

CHAPTER XI

THE COLOURS OF AMPHIBIANS AND REPTILES

CHAPTER XII

THE COLOURS OF BIRDS

CHAPTER XIII

THE COLOURS OF BIRDS (*Continued*)

CHAPTER XIV

THE COLOURS OF MAMMALS AND THE ORIGIN OF PIGMENTS

CHAPTER XV

THE RELATION OF FACTS TO THEORIES

INTRODUCTION

To those who have not followed closely recent developments of Evolution Theory, the connection between Biology and Colour may seem very remote. The phenomena of colour, it may be said, are entirely the province of the physicist; that the sky is blue and the grass green are two facts of similar nature, and the one is as inexplicable as the other. So in general it may be said that it is simply a fact of experience that most objects, whether animate or inanimate, present themselves to our eyes as coloured, and that it is therefore absurd to separate the phenomena of colour as they appear in organisms from the similar phenomena of inorganic nature. A little reflection will, however, convince every one that the biologist cannot afford to be indifferent to the colours of the organisms with which he has to deal. In the first place, they attract his attention because of their frequently great intrinsic beauty and their arrangement into patterns and markings which may

B

exhibit extraordinary constancy. Such markings often tend to reappear in slightly modified forms in a large series of nearly related organisms. Thus Eimer has shown that the markings of the head, often so conspicuous in the domestic cat, tend to recur constantly throughout the whole of the Carnivora, and in a large number of cases may be quite definitely traced. This constancy of colour or marking is not infrequently available for the purpose of classification, or at least of ready identification, and has therefore always attracted much attention.

But the point of interest about the colours of organisms which has of late had most stress laid upon it is one which, like so many recent developments of biological theory, has grown out of Darwin's work. As is well known, Darwin's statement of the Doctrine of Descent involved a clear formulation of what has become widely known as the Struggle for Existence, but what was in fact the first clear appreciation of the intricacy of the relations existing between organisms and their environment, the term including both physical nature and other organisms. Darwin endeavoured to prove that the balance of nature is so finely adjusted that the slightest oscillation of one part may affect parts apparently far removed from it, and that the Struggle for Existence is so keen that all specific characters are, as it were, maintained at the point of the sword. Now we have just seen that the colours and markings have always been recognised as among the most constant of the characteristics of organisms ; if, therefore, all specific characters are preserved by virtue of their usefulness, then surely the colours must be of supreme importance to the organisms—whence we have all the recent developments of colour theories. For a historical sketch of the way in which many theories as to the meaning of colour in organisms have been built up around the central doctrine of Natural Selection, reference should be made to Mr. Wallace's *Darwinism*.

It is, however, of some importance to note that the interest in the phenomena of colour thus manifested has been very largely polemical in nature. The supreme test of the value of a new scientific hypothesis is that it explains series of phenomena hitherto inexplicable, and the eagerness with which the colours of organisms have been discussed and investigated has been chiefly stimulated by the hope of adding fresh arguments to those already brought forward in support of Darwin's theory. We do not propose at present to discuss the question of how far this hope has been realised, but shall rather pass on to consider a third aspect of the colours of animals which has been as yet little dwelt upon. This is the meaning of the colour in the functional economy of the organism. The problems of colour have of late years been so exclusively considered from the outside—in connection with the relation of the organism to its environment—that we are apt to forget that the colours must also have a meaning and a justification in the physiological processes of the individual. The present method of treating the subject is in large part the result of that habit of regarding organisms as mere bundles of qualities which has been so keenly criticised by a recent writer. If we once realise that the colour of an organism is not an isolated characteristic forced upon it, as it were, from without, but may be merely the outward expression of its constitution, we may surely hope not only to be delivered from many laborious hypotheses as to the use of colour in particular cases, but also may perhaps learn something of the physiology of colour. The subject is at least now sufficiently important to merit consideration from a purely physiological standpoint.

We thus see that there are three reasons why it is desirable that the Biologist should concern himself with Colour in organisms. The first is the conspicuousness of colour phenomena in a merely objective survey of animals and plants ; the second is the relation of these colours to current theories of evolution ; and the third is their importance in comparative physiology.

The order of arrangement of these three is not purely formal, but to some extent corresponds to the historical order in which the subject has been studied. Before Darwin, the colours of organisms were chiefly studied as convenient marks by means of which the organisms could be recognised. When Darwin put forward his theory, which is based on the supposition that all specific characters are of supreme importance, it was a natural deduction from the constancy of many colours that these must be of great importance, and so we have all the modern theories of colour (see Mr. Poulton's *Colours of Animals*). Again, now that the theory of Natural Selection is no longer the centre of men's thoughts, and search is being made for a deeper analysis, it is recognised that it is probable that the phenomena of colour have some connection with physiology, or are at least not wholly accounted for when their usefulness is proved (see Mr. Beddard's *Animal Coloration*, and, for the general question, Mr. Bateson's great work on Variation).

Having now justified the intrusion of the biologist into what appears to be the domain of the physicist, we are at liberty to face the phenomena of colour as they appear in organisms. As, however, the relation of colour phenomena to the theory of Natural Selection has had so much attention bestowed upon it, we do not propose, at least in the first place, to consider it here, but rather to direct attention to the chemical, and where possible, to the physiological aspects of the colours and pigments of organisms. In a final summary it will be necessary to consider the bearing of the facts upon theories.

CHAPTER I

THE COLOURS OF ORGANISMS

Colour Phenomena in General—Distinctions between Pigmental and Structural Colours—Characters of Structural Colours and their Classification—Production of Light by Organisms (Phosphorescence).

IT is not necessary here to consider in detail the physical aspect of colour. Every one is more or less familiar with the fact that, in general terms, objects appear to us to be coloured because they absorb certain of the elements of white light and reflect the remainder. Thus the substance vermilion appears to us to be red because it absorbs all the components of the incident light except the red, and this it reflects ; it is a fundamental property of vermilion that it possesses this power of selective absorption. Practically all bodies do exercise selective absorption to a greater or less extent, but it is only when the absorption is marked in the visible part of the spectrum that we definitely recognise the bodies as coloured. When such coloured bodies can be employed to impart their own colour to animal or vegetable substances they are known in the arts as pigments, but to the biologist the term includes only substances which are produced by the activities of plants or animals. The colour of the rose, for example, is due to a red pigment present in the cells of the petals, which can be extracted from those cells. This is the simplest form of colour-production, and is the one which commonly occurs in plants. Colours so produced are called pigmental colours and can be recognised by the following characters. Pigmental colours are those produced by pigments, or substances of definite chemical composition, which can by appropriate re-agents be extracted from the coloured tissues, and which react to light in the same way whether they are within the tissues or outside of them. Tissues or organisms showing only pigmental colours never have a surface gloss, and the colour is not altered by immersion in any medium which does not directly attack the pigment.

Although the great majority of the colours of plants are produced in this simple way, yet even among them we have indications of that other kind of colour-production which reaches its climax of splendour in birds and butterflies. The gorgeous tints of the humming-birds, which during life change with every movement, are not produced by the dyeing of the feathers with pigment, but are phenomena of the same order as the colours of the gems after which some of the birds are named. We all know that the colours of the opal and of many minerals are not due simply to a prime property of the substance concerned, but are optical effects dependent upon various external conditions for their full manifestation. Many of the most brilliant tints of animals are similarly optical colours, but as they are produced by the special structure of the coloured parts, they are more frequently known as structural colours.

Structural or optical colours in organisms can be recognised by the following tests. They usually vary either with the angle of incidence of the light, or with the change from reflected to transmitted light ; tissues or organisms showing these colours have a very marked surface gloss, and the colour is usually destroyed by injury to this surface ; immersion in a neutral medium whose refractive index is different from that of air, also usually destroys the colour. Optical colours may further be recognised by the negative character that the coloured tissues do not yield to any reagent a pigment of the same tint as that which they themselves possess. A peacock's feather affords an excellent example of a type of structural coloration which is widely spread in birds. If the reader stand in front of a window and hold a peacock's feather in his hand nearly on a level with the eye, and then, still with the feather at arm's length, slowly describe a semicircle on his own axis, he will note that the colours of the feather undergo a complete cycle of changes. In this case, therefore, the colours change with the change of the angle of incidence of reflected light. If the feather be now held up to the light it will be seen that the colours disappear, to be replaced by a dull brown or black tint. Thus the feather changes colour according as it is viewed by reflected or transmitted light. The presence of a marked surface gloss is, of course, very obvious, and, though the size of the coloured particles is too small to make it easy to note the effect produced by injury to this surface, yet the disappearance of the colour when the surface is thoroughly wetted with water or oil can be observed very readily.

Having now distinguished generally between the colours due to pigment and those which are the result of a special structure, we must proceed to consider these different methods of colour-production in detail. As much less is known of the structural colours than of the pigmental, we shall devote the remainder of this chapter to the former, leaving the latter for a further chapter.

CHARACTERS OF STRUCTURAL COLOURS AND THEIR CLASSIFICATION

1. *White* is probably the simplest and most readily understood of all structural colours, and, except in rare cases, is always structural. The colours of the lily, of white feathers, of Arctic mammals, are all due to the same cause; namely, the total reflection of light produced by the intercalation of numerous bubbles of air, or some other gas, among colourless solid particles. Thus in the lily the colourless cells are separated by very numerous intercellular spaces containing air. This is one of the simplest forms of structural colour, not only because it is so readily explicable physically, but also because there is no complication arising from the presence of pigment in the tissues, and because the modification of " structure " necessary to produce it is of the simplest nature.

Most people are familiar with the analogous process of producing a white colour by pounding up colourless glass, or crystals of blue sulphate of copper, while the whiteness of snow, which has furnished so many metaphors, is produced in a precisely similar manner.

The fact that the whiteness of all these substances is due to an optical effect and not to a pigment should be thoroughly grasped, otherwise those not accustomed to dealing with colour phenomena will find much difficulty in comprehending structural colours in general. White sunlight is produced by the combination of all the tints of the rainbow. When objects permit light to pass completely through them, we call them transparent; when they reflect all the rays of the light uniformly, we call them white. This whiteness may be produced in one of two ways. A substance such as "Chinese white" is white because it is a property of the particles of which it is composed to reflect equally all the rays of incident light; it is further a familiar fact that Chinese white can be employed to impart its own colour to other objects, that is, it can be employed as a pigment. Snow, on the other hand, is white, not because its individual particles reflect the light—on the contrary they are transparent—but because these transparent particles are separated by bubbles of air. The incident light in passing from the one medium to the other is bent or refracted, and the result is the appearance of whiteness. A white colour in organisms, except in very few cases, is similarly produced, and is not due to pigment.

Other structural colours are to be accounted for

in a closely similar fashion. Glass is a colourless or transparent substance, when powdered it is white, when cut with prismatic edges it displays all the colours of the rainbow, and yet the qualities of the glass remain unaltered. These are of course very familiar facts, but it is important to realise also that many of the most brilliant colours of organisms are produced in a similar fashion—are adventitious and not due to the essential properties of the coloured substance. It is only in rare cases, however, that bright colours are produced in a way capable of simple physical explanation. There are usually complications arising from the presence of some amount of pigment, from the superposition of tissues, or from the complex nature of the individual tissues. Such brilliant optical colours occur apparently only in cuticular structures. It is a curious fact that, although such structures are of course not cellular, nor living, yet their colour frequently fades very rapidly after death ; that dragon-flies, for example, lose in a very short time all their gorgeous tints is a fact only too well known to collectors. This may be due to loss of water, or to changes in the underlying tissues.

Structural colours are most brilliant and conspicuous in birds and insects, but it is chiefly in the former that they have been studied. Dr. Gadow has especially studied the structural colours[1] of birds, and he divides them into two classes, according to their behaviour as regards incident light. Thus certain structural colours, such as green and blue, are unchanging in reflected light, and are then

[1] For a purely physical treatment see a pamphlet by B. Walter.

not readily distinguishable from pigmental colours. These Gadow classifies as *Objective Structural* colours. Again many colours change in tint according to the angle at which they are viewed. Such metallic colours may be classified as *Subjective Structural* colours.

2. Of *Objective Structural* colours, green and blue afford the best examples. Green seems to be usually produced by a combination of a yellow pigment and a structural modification, wherefore green feathers usually appear yellowish in transmitted light. The display of a blue colour again seems, at least in birds and probably in insects, to be always associated with the presence in the tissues of a dark-coloured pigment. A blue colour in the feathers of birds is always confined to certain parts of the feather, and its presence is associated with considerable modifications of feather structure. Blue is not however confined to the feathers, but may occur also on bare patches of skin, as in some of the Paradise birds, in the cassowary, etc. Curiously enough, a blue colour in the skin seems to be particularly unstable, fading as rapidly after death as does the blue colour in the abdomen of some dragon-flies.

The exact physical cause of green and blue colours is still doubtful, but the fact that the colours are structural is readily perceived both by the strong surface gloss and by the disappearance of the colour in transmitted light.

3. While unvarying blue and green tints in birds rather heighten the general effect than display in themselves great beauty, the *Subjective Structural* colours, on the other hand, display the most ex-

quisitely varying tints, and it is to them that many birds and insects owe their wonderful flashing beauty. These colours glow with all the tints of the rainbow, and change with every changing ray of light. Such metallic colours are of course not uncommon phenomena in the inorganic world, and are displayed, for example, with extraordinary brilliancy on a polished slab of the mineral Labradorite; among organisms, although suggested elsewhere, they attain their maximum brilliancy in birds and insects. In birds their degree of development varies greatly, for we find them ranging from the dull greenish gloss of some of the female humming-birds to the gorgeous colouring of many of the males in humming-birds and birds of Paradise. Among birds metallic colour is apparently always associated with the presence in the feathers of dull brown or black pigments, which are necessary for the production of the colours. It is also associated with a modification of the feather structure which in many cases renders the feathers unfitted for the purpose of flight. Here, as in the case of objective structural colours, the exact physical causation of the colours is unknown.

In insects the colours are equally bright, but have had little attention bestowed upon them; it is still doubtful whether the colours in them are or are not associated with the presence of dark pigment.

4. The presence of a pigment is not, however, essential to the production of structural colours; the common earthworm, for example, exhibits a faint iridescence which is due to the presence of numerous fine lines on its colourless cuticle, these fine lines producing interference of light. Although the colour-

ing is very slight in the earthworm, it is well known
that some of the marine worms, *e.g.* the sea-mouse
(*Aphrodite*), are covered with numerous bristles which
exhibit brilliant iridescent colours.   Again, the colours
of mother-of-pearl are of course produced by structure
only, without any assistance from pigment.   For convenience of reference, Structural Colours may perhaps
be arranged as follows, retaining Gadow's distinction
of objective and subjective colours.

Structural Colours :—

1. Those not dependent upon the presence of a
   pigment.
   (*a*) Due to total reflection ; white colour of
        some flowers, of some feathers, of hair, etc.
   (*β*) Due to striation of the surface, occurrence
        of thin plates, etc. ; iridescence of bristles
        and cuticle of worms, of mother-of-pearl, etc.
2. Those dependent on the presence of a pigment.
   (*a*) Objective structural colours ; blue and
        green feathers.
   (*β*) Subjective structural colours ; metallic
        colours of many birds and insects.

Structural colours are of extreme interest, not
only on account of their wonderful beauty, but also
on account of the difficult questions connected with
their origin.   It is to some extent possible to correlate pigment-production with the physiology of the
organism, but this seems extremely difficult in the
case of structural coloration.   We may note, however, that structural colouring attains its greatest
perfection among birds and butterflies, and both
groups are noted for the extraordinary development
of their cuticular structures.   The delicate beauty of

the sculpturing of butterflies' scales has been extolled by most possessors of a microscope, while savage and civilised races are alike in their admiration for the feathers of birds. The fact that organisms so widely separated as are birds and butterflies are alike in exhibiting both exquisite structural coloration and a wonderful development of structures arising from the cuticle, suggests that the structural colours are in origin merely a result of extreme differentiation of the cuticle, and therefore produced by the same cause which gave rise to this differentiation. The presence of brilliant iridescence in some of the mud-inhabiting worms is therefore not quite inexplicable, for here also we find that the cuticle shows a considerable amount of differentiation. We can also further understand how it is that the highest pitch of perfection is attained in birds and butterflies, when we consider that in both cases the colouring occurs in connection with structures which are of supreme importance to the species, that is, with the feathers of the bird and the scales, which are but outgrowths of the wings, in butterflies.

PRODUCTION OF LIGHT BY ORGANISMS (PHOSPHORESCENCE), ITS DISTRIBUTION AND MEANING

The production of light, the phenomenon commonly though incorrectly spoken of as phosphorescence, is so striking a characteristic of many organisms that it deserves notice here in our consideration of colours. It is unnecessary to enter into a detailed account of the theories of Dr. Carpenter and others as to the probable use of this characteristic, because these are already familiar to most people. It is more to our purpose to note the organisms among which it occurs, and the special parts with which it is associated.

To begin with the simplest organisms—the fact that many decaying substances shine in the dark has long been known, and modern bacteriologists have been able to isolate the specific micro-organisms which possess, under certain conditions, this property of evolving light. Just as in the case of pigment production, so also the phosphorescence depends upon the nature of the organism and the character of its surroundings. Thus *Photo-bacterium phosphorescens* evolves light during the fermentation of sugar if free oxygen be present and the temperature be favourable (3°-35° C.).

Among the Protozoa, phosphorescent forms are most noticeably represented by the genus *Noctiluca*, but some Radiolarians, such as *Thalassicola*, are also luminous. In *Noctiluca* the phosphorescence is said by Allman to be produced in the cortical layer of protoplasm.

Among multicellular plants phosphorescence is apparently confined to the fungi. It is an old controversy whether it does or does not occur in flowering plants, but most authorities answer the question in the negative. In the fungi the luminosity is marked in several species of *Agaricus*. The phenomenon is only manifest in the presence of free oxygen, is associated with oxidative processes, and is dependent upon the temperature. The light is said to be usually of blue or greenish colour, but in some cases is quite white.

In the multicellular animals it is curious to note hat phosphorescence is most marked in pelagic and n abyssal animals, though also of course occurring lsewhere, *e.g.* in the glowworm, some centipedes, etc.

In the Cœlentera, many of the Medusæ are especially remarkable for their luminosity; their colouring is thus described by A. Agassiz, "The jellyfishes, sparkling and brilliant in the sunshine, have a still lovelier light of their own at night. They send out a greenish-golden light, as lustrous as that of the brightest glowworm, and in a calm summer night the water, if you but dip your hand into it, breaks into shining drops beneath your touch."

As to the remaining groups of the Cœlentera, we find well-marked luminosity in the Siphonophora, in both divisions of the Anthozoa, and in the Ctenophora. In the first group we may specially note *Physalia*, the Portuguese Man-of-War, which Agassiz describes as appearing like a fire-balloon at night. Of the Anthozoa, the Alcyonaria include the most brilliantly phosphorescent forms; the Gorgonidæ being often described as forming luminous forests at the bottom of the sea, where those of the deep-sea forms which retain their eyes are supposed to congregate. Luminous forms occur also, however, among the Zoantharia, where the deep-sea *Sagartia abyssicola*, for example, is said to secrete abundant phosphorescent mucus.

In the Ctenophora phosphorescence is so common as to be practically a class character, and the members of the group are frequently very important in the production of surface phosphorescence.

In Worms certain of the marine Annelids, such as *Chætopterus* and the Syllidæ, are luminous.

C

Among Echinoderms phosphorescence seems to be known only among the Ophiuroidea, where it has been described both in Mediterranean and in Atlantic species.

Among the Crustacea phosphorescence is exceedingly common, especially among deep-sea forms. In some cases the luminosity is confined to the eyes, in others, as in many of the deep-sea Schizopoda, there are special luminous organs placed behind the eyes or above the legs, in other cases again the phosphorescence is diffused.

In the terrestrial Anthropods phosphorescence is known among various insects, glowworms, fireflies, etc., and among centipedes, but it is not so common as among marine forms.

Among the Mollusca phosphorescence occurs among pelagic forms such as *Phyllirhoë*, and among burrowing forms such as *Pholas*, whose luminosity has long been known.

Next to the Cœlentera the Tunicata are perhaps the forms which display the most universal and brilliant phosphorescence. Every one knows Moseley's description of the brilliancy of the light emitted by *Pyrosoma*, the phosphorescent fire-flame. Many other pelagic Tunicates display the same phenomenon to a lesser degree. The Tunicates display so many striking analogies to the Cœlentera that this physiological one is perhaps not remarkable.

Finally, we have the phenomena of phosphorescence admirably displayed in fishes, especially the pelagic and abyssal forms. In the pelagic forms the sense-organs and the lateral line are often phosphorescent, while in some cases there are special luminous organs. Many of the deep-sea fishes are very markedly phosphorescent ; the phosphorescence may be limited to spots on various parts of the body surface, or there may be special luminous organs on the head, while in many cases there are delicate tactile processes phosphorescent at the tip ; in some cases the fins are themselves luminous.

As to the details of the mechanism of phosphorescence, Prof. Panceri of Naples made some interesting observations more than twenty years ago. He studied many marine animals, especially *Pyrosoma*, *Phyllirhoë*, *Pholas*, and others. His papers display a combination of careful observation and acute deduction which make them models of scientific investigation. By the employment of very ingenious methods he demonstrated the fact that in *Pyrosoma* the luminosity is due to two cell-clusters in each ascidiozooid. As each colony contains thousands of individuals, the number of luminous spots is enormous even in a relatively small colony. The cell-clusters lie " on each side of the anterior end of the branchial sac," not far from the nerve ganglion, and consist of spherical glandular cells. They contain a substance soluble in ether, probably of fatty nature, which apparently becomes luminous when oxidised. In ordinary circumstances the light is only excited by a mechanical stimulus, and then is first aroused at the stimulated point and spreads more slowly throughout the colony. These " luminous currents " are not nearly so rapid as in some of the Cœlentera, *Pennatula* for example, and to this is probably due the fact that when Moseley wrote his name on a large *Pyrosoma* it stood out in " letters of fire," which of course could not occur if the propagation of the stimulus were rapid. Under ordinary conditions the phosphorescence is stimulated by the shock of the waves against the sensitive organism, but it may be also produced by various chemical stimuli, such as fresh water, alcohol, and ether. The action of the two latter agents is very interesting. If they come into direct contact with the luminous organs they extinguish the light instantly, while if they do not reach the organs they act as powerful stimulants of the light. The luminosity ceases at death, but the fluid obtained from the living organism by crushing retains its phosphorescent power for some time.

In the Medusæ the luminosity is due to the epithelial cells, and in consequence of the delicacy of these the light is communicated to surrounding objects. In *Phyllirhoë* the light is produced by the ganglion cells of the nervous system, and by certain peculiar nerve cells (Müller's cells), and is again apparently due to a special substance occurring in the luminous cells.

In the Copepoda the phosphorescent substance seems to be produced in special glands situated at various parts of the body, and the light, according to Giesbrecht, is only produced when the substance secreted comes into contact with sea-water. In *Metridia longa*, according to Vanhoeffen, the light appears especially immediately behind the head, and at a point close to the posterior end of the abdomen. The living organism examined in light is colourless, except at these same points which appear as "moss-green" spots; these spots are apparently the phosphorescent glands. Other phosphorescent Copepods display similar phenomena.

We cannot here enter into a description of the phenomena of phosphorescence as they appear in all luminous organisms, but may conveniently add to the descriptions already given, an account of the process in terrestrial forms. For this purpose we may take Prof. C. Emery's account of the phosphorescence of *Luciola italica*, the Italian glow-insect. Here in the male the luminous organs are placed on the two last abdominal segments, each one appearing as a continuous surface although originally formed in two halves. In the female there are two luminous spots at the sides of the ventral surface of the fifth abdominal segment, the following two (the last) abdominal segments appear in dried specimens of the same pale colour as the fifth but are not luminous. The luminous organs consist each of a double layer, a ventral transparent one and a more dorsal 'chalk-white one containing masses of urates. Some of the large tracheal tubes run along the inner surface of the dorsal plate, and give off perpendicular branches which pass downwards and penetrate into the substance of the transparent layer, where they branch very freely, and come to lie close beneath the skin.

A surface view of the luminous organ shows a series of round or oval spots corresponding to little masses of transparent cells surrounding the termination of these tracheal branches; it is these spots which are luminous during life. They turn brown when treated with osmic acid, and are separated by broad darker intervals. More careful examination shows that these spots correspond to little cylinders formed by clusters of cells—the tracheal end-cells—

which are placed near the ends of the fine tracheal capillaries. The substance of the luminous organ is made up of parenchyma cells, most easily perceived in the transparent layer. The result of his examination convinced Emery that the luminous organ was a specialised portion of the fatty body, the arrangement of the tracheæ being one of the most marked indications of specialisation.

His explanation of the luminosity is as follows: the parenchyma cells (fatty cells) secrete the luminous substance, from them it is taken up by the tracheal end-cells, and, being here exposed to the action of the oxygen of the tracheal system, undergoes a process of oxidation resulting in the formation of light. This process can only occur in regions where the chitinous lining of the tracheæ is very thin, as in the capillaries ; the free branching of the slender tracheal capillaries in the luminous organs especially provides for this. In a later paper Emery is inclined to lay more stress upon the parenchyma cells than upon the tracheal end-cells as the seat of luminosity. The luminosity is then due to the oxidation (or combustion) of a probably useless body stored in the cells of the organ (the fatty body). To this description we may perhaps add that recent investigations tend to emphasise the importance of the fatty body of insects as an organ connected with excretion.

As to the nature of the light emitted by the different phosphorescent organisms, the observations are not sufficiently numerous to draw any conclusions. Moseley found that the blue and violet rays were absent in the light of three deep-sea Alcyonarians, while surface forms displayed light of various colours ;

but the observations are too limited for any conclusion to be drawn.

With regard to phosphorescence in general it
appears to be most common in the relatively simple
organisms of the pelagic fauna, and next to these in
the abyssal fauna. If, as there seems much reason
to believe, it depends upon processes of oxidation,
its presence in the deep-sea forms is very remarkable.
The phosphorescent organs of the deep-sea forms,
especially the fishes, occur most frequently in connection with nervous organs, especially with those
delicate tactile processes so characteristic of these
animals. A good example of this is found in the
luminous " lures " of the deep-sea fishing frog. The
statement that this form has adopted the salmon
poacher's method and does its fishing by means of a
luminous lure, seems at first sight to demand a teleo-
logical explanation, but when we remember that the
lures are greatly developed tactile organs the fact
that they are luminous is less inexplicable, in origin
at least. The whole subject of the phosphorescence
of abyssal forms is exceedingly puzzling. We are all
familiar with the pictures of the abysses of the ocean,
of the chill darkness illuminated by fitful gleams of
phosphorescent light, but it is difficult to decide how
far these representations are justifiable. It is a
somewhat obvious remark that no one has seen this
lower world, but is one which is perhaps not altogether
unnecessary. Many deep-sea organisms phosphoresce
when brought to the surface, but there is no *proof*
that they do so under the ordinary conditions of their
life. Has the increased amount of oxygen in the
surface waters no intensifying effect? Then also the

Alcyonarians have been described as forming luminous forests in the ocean depths, but Agassiz, Panceri, and others insist that in the cases studied by them, the luminosity is dependent upon mechanical and chemical stimuli, and we are always being told of the calm stillness of the ocean abysses. The swimming of fishes among these forests may set up currents which produce luminous flashes, but surely those elaborate hypotheses which assume that the Alcyonarians give sufficient light for the other organisms to see each other's colours by, are a "baseless fabric" of the imagination.

With regard to the phosphorescence of the deep-sea fishes, it is interesting to note that Agassiz considers it to be a characteristic of these abyssal fishes which belong to groups, most of whose members are pelagic, rather than of abyssal fishes in general.

As to the meaning of phosphorescence little can be said. It appears not improbable that like pigments it may arise in different ways in different organisms. It is at least difficult to believe that a process like that described by Emery for *Luciola*, which seems to be purely excretory, can be entirely homologous with the luminosity of nervous structures. The luminous substance in all cases is apparently of the nature of a fat, and it is unnecessary to emphasise the fact that complex fatty substances tend to occur in association with nerve tissues.

The subject has an important bearing upon the general one, for phosphorescence, no less than colour, has been frequently dismissed with a few magic words upon Natural Selection.

CHAPTER II

THE PIGMENTS OF ORGANISMS

Natural and Artificial Pigments—Classification of Pigments—
Pigments of Direct Physiological Importance—Derived
Pigments—Waste Products—Reserve Products—Intro-
duced Pigments—Distribution of Pigmental Colours—
Spectroscopic Characters of Pigments.

ALTHOUGH, as we have seen, some of the most
beautiful colours of organisms are optical effects, or
are produced by a combination of a pigment and a
certain structure, yet the vast majority of the colours
of plants, and the colours of many animals, owe their
origin to pigments only. Before proceeding to consider the colours of the separate groups of animals
and plants, we shall look for a little at pigments in
general.

The pigments or colouring-matters which are
most familiar to artists and those engaged in the
industries, are very frequently either artificial organic
substances, such as the now familiar aniline dyes,
or are inorganic compounds of the metals, such as
ultramarine and Scheele's green. Some, however,
of the natural organic pigments are used in the arts.
As is well known, the ancients obtained their Tyrian

purple from various small shell-fish, such as *Murex*
and *Purpura*, while the different preparations of
carmine and cochineal, saffron, and hæmatoxylin or
logwood are other familiar examples of pigments
produced by animals or plants which are habitually
employed as colouring-agents. The greater number
of the pigments of plants and animals are, however,
too fugitive to be so utilised, and in consequence are
in their pure state unfamiliar to most people. Such
are chlorophyll itself, the blue, red, orange and
yellow pigments of flowers, and the green, red,
yellow and orange pigments of animals, most of
which are destroyed by light or other agents with
great rapidity. It is indeed worth noting that most
of the natural dyes are of vegetable origin, and that
they are usually inconspicuous and apparently unimportant during the life of the plant, cf. hæmatoxylin, brasilin, etc. So, too, when they are of
animal origin, they are not usually important in
producing the coloration of the living creature, cf.
Murex, the coccus insect, etc. It is partly in consequence of the fact that the pigments important
in producing coloration are so fugitive and so
difficult to obtain in a pure state, that so little is
really known of their relations and constitution.
Many of them, indeed, have hardly been investigated
at all.

We may note here a point which will be dwelt
on in detail in its proper place, that a pigment
found in an animal need not necessarily be formed
by that animal. There are some instances of
pigments which are transferred from one animal
or plant to the tissues of another by the food with

apparently little or no change in composition. A more remarkable case still is found in the bones of certain fishes (*Belone, Protopterus, Lepidosiren*), of an amphibian, and of a lizard. In these the bones are coloured a bright green and have been found to contain vivianite, a mineral consisting of a phosphate of iron. This mineral has also been found in connection with fossil bones, but is there probably the result of a chemical change during the process of fossilisation.

In the vast majority of cases, however, the pigments of plants and animals are definite organic compounds, produced by the organism in which they occur. Such pigments are of very varied characters and composition, some being substances of great complexity, while others are relatively simple. Just as in the laboratory of the chemist, a coloured substance capable of being employed as a pigment may be produced at many stages in a series of chemical transformations, so in the laboratory of nature pigments may occur at many stages in metabolism. This at once suggests an interesting question—is there any relation between the colour and the chemical composition of pigments? That is, are the pigments of simple chemical composition usually certain particular colours, and those of complex structure other colours? Such statements have under various forms been made repeatedly, and later we must consider them in detail, but there is as yet no evidence to show that the colour of pigments can be employed as a direct means of classification.

The classification of the pigments of plants and animals is indeed a matter of profound difficulty.

The chemistry of most is so imperfectly known that the composition cannot be employed as a basis, while the suggestions as to physiological function, with which the literature of the subject is full, are for the most part merely guesses.　But while we admit that it is at present impossible to draw up a logical classification, it may be well to mention certain categories into which some of the better known pigments are said to fall.　These are as follows :—

Native Pigments.	1.	Pigments of direct physiological importance, as in respiration, etc.
	2.	Derivatives of such pigments.
	3.	Waste products or modifications of such.
	4.	Reserve products or pigments associated with reserves.
	5.	Introduced pigments.

The pigments falling under these heads we shall consider in some detail.

1. PIGMENTS OF DIRECT PHYSIOLOGICAL IMPORTANCE

The pigments in this group have always attracted much attention, and have been attentively studied in many cases.　As types we may take hæmoglobin and chlorophyll, the one so characteristic of higher animals, the other of plants.

Hæmoglobin.—As is well known, hæmoglobin, the red colouring-matter of the blood of Vertebrates, is a compound of a pigment containing iron (hæmatin) and a proteid, which is one of the few as yet obtained in crystalline condition.　Its great im-portance is due to the fact that, owing probably to

the contained iron, it possesses the power of forming
a loose combination with oxygen, and so of acting as
a respiratory pigment.   Of its origin little is known,
but there seems to be some reason to believe that in
development it arises from the chromatin of the
nucleus.   In a recent paper on iron-compounds in
animal and vegetable cells by Dr. Macallum, the
author puts forward the theory that the chromatin,
that part of the nucleus which readily takes up stains,
is an iron-holding nucleo-albumen.   This chromatin
has the power of fixing free oxygen, and the formation of hæmoglobin results in the conversion of the
nuclein into the pigment hæmatin ; the substance retaining its primitive power of forming a combination
with oxygen.   Further, Dr. Macallum regards the
diminished amount of hæmoglobin in the blood in
anæmia not as the prime cause of the disease, but as
one of the results of the deficiency of iron-containing
compounds in the nuclear chromatin ; this deficiency
being the prime cause of the various symptoms of the
disease.   The distinction may seem unimportant,
but it is not so in reality ; for if Macallum's view be
correct, it carries us one point further.   Anæmia is
recognised in our own species by the pallor of the
lips and skin produced by the diminution in the
number of red blood corpuscles ; it is therefore
natural to conclude that the disease is caused by
the insufficient amount of hæmoglobin present.   If,
however, it is proved that this absence of hæmoglobin
is itself merely a consequence of something deeper
which affects the whole organism, we are at least a
little nearer the problem of the primary meaning of
hæmoglobin.

So far we have spoken of hæmoglobin as the supremely important pigment of Vertebrates, as a pigment perpetually justified by its usefulness, but we must notice that it has a wide distribution in the animal kingdom and occurs under circumstances where usefulness is difficult to prove, although this has not prevented the assertion being widely made.

In the striped muscles of Vertebrates, for example, it is widely but irregularly distributed, often in a single species being invariably absent in some muscles, and invariably present in others. Of a well-marked distinction between red and white flesh the rabbit is a familiar example; the common fowl is another as well known to the physiologist as to the epicure. Other notable cases are those of the fish *Hippocampus*, where only the muscles of the dorsal fin are red, and of the rare fish *Luvarus*, where the difference between red and pale muscles is very well marked; but it would be easy to multiply examples almost indefinitely. Among the unstriped muscles of Vertebrates, hæmoglobin is said to be found only in the wall of the rectum.

Among Invertebrates, hæmoglobin shows the same peculiarities of distribution as in the muscles of the Vertebrates. Thus it is present in the perivisceral fluid of some Turbellarians, of *Glycera*, and of *Phoronis*; in the hæmolymph of *Lumbricus, Tubifex*, and other Annelids; in the muscles of the pharynx in *Buccinum undatum, Littorina*, and other Gasteropods; in the sheath of the nerve-cord in *Aphrodite aculeata*; in the cephalic slits of Nemerteans; and so on.

It is well known that in the case of the red

and the pale muscles of Vertebrates the difference of colour is usually associated with differences in chemical composition and histological character, and in certain cases at least with physiological differences in relation to the reaction to stimuli. It is therefore sometimes supposed that the contained hæmoglobin is of respiratory importance, and is the cause of the physiological characters. On the other hand, in certain cases as in the rabbit, the association between the colour and the other characters is not very close, and it is to be noted that insects also possess two kinds of muscle distinguished *inter alia* by their colour ; in their case the extensive development of the tracheal system forbids the idea that the difference is caused by the presence or absence of a special respiratory pigment.

In Invertebrates the difficulties connected with the supposition that hæmoglobin is always of supreme importance as a respiratory pigment are even greater.

It is said that hæmoglobin is especially necessary to *Lumbricus* on account of its peculiar habitat ; that its presence in the head-slits of Nemerteans is essential for the oxidation of the brain ; that it is present in the muscles of the buccal mass of *Littorina* because these muscles are especially active ; and so on. On the other hand, many large marine worms have no hæmoglobin, whatever their habitat, and many Gasteropods have none in their buccal muscles. Can we suppose that these muscles are less active in the limpet, the snail, and many others than in *Littorina* ?

The more these questions are considered, the more difficult does it become to suppose that hæmoglobin in these cases is functional in a degree at all comparable to that seen in Vertebrates. If it were, we should expect that when once acquired by the members of a group it would be retained by all their descendants, however widely they might diverge in other respects ; and the irregular distribution of hæmoglobin among the Invertebrate groups is contrary to this supposition.

If, however, we accept Macallum's theory that hæmoglobin is a modification of nuclear chromatin, then it may quite well be that it has a primary meaning in metabolism which accounts for its presence in Invertebrates and in the muscles of Vertebrates, while under certain conditions, as in the case of the hæmoglobin of the blood of almost all Vertebrates, it may acquire supreme importance as a respiratory pigment. Such a suggestion would explain many peculiarities of distribution which are at present exceedingly puzzling.

Hæmoglobin is of course a chemical compound of great complexity. Just as it is itself probably formed by the modification of a still more complex substance, so it in its turn undergoes a process of breaking down, during which it gives origin to a number of simpler substances. Some of these form pigments which will have to be considered under Group 2.

Chlorophyll.—Chlorophyll is, like hæmoglobin, a compound of considerable complexity, but owing to its great instability its exact composition is unknown. We have, further, no exact knowledge as to the part it plays in the metabolism of the plant. It is of course a familiar fact of experience that green plants when exposed to sunshine can decompose carbonic acid and set free oxygen, and that it is by virtue of their chlorophyll that they can do this, but beyond this we have nothing but a host of rival theories. The most interesting of these is perhaps that of Schunck, who considers that chlorophyll carries carbonic acid from the air to the assimilating protoplasm, just as hæmoglobin carries oxygen. Macallum holds that chlorosis in plants and animals is due to the same cause, the absence or deficiency of iron in the nuclear chromatin ; and this in spite of the fact that it is improbable that chlorophyll contains iron. Thus the absence of chlorophyll in a plant grown without iron is only one of the consequences of the general unhealthiness. In reference to this view of Macallum's, it is interesting to note that according to Nencki there is a close relation between hæmatoporphyrin, a derivative of hæmoglobin, and one of the derivatives of chlorophyll green.

As is well known, chlorophyll is, at any rate in solutions, an extremely unstable pigment, fading in light with great rapidity. In natural conditions it is associated with one or more pigments belonging to a widely spread group of pigments—the lipochromes, characterised by their colour, which varies from red to yellow, as well as by other properties. It is uncertain whether these subsidiary pigments exercise any influence in the process of assimilation, or whether it is not the true chlorophyll or chlorophyll green which is alone active. Lipochromes are not the only pigments which occur in association with chlorophyll, but the subject will be treated again in further detail when we come to consider the colours of plants.

So far as is yet certainly known, chlorophyll is the only pigment by means of which organisms can avail themselves of the carbon of the atmosphere, though it is quite possible that some other pigments may yet be shown to possess this property. The suggestion has been made in the case of the bright pigments of coral polypes, as well as of various other pigments, notably of the purple pigment of *Beggiatoa roscopersicina* (see Chap. III.).

Pigments fulfilling the function of hæmoglobin are perhaps fairly numerous, but there is some doubt connected with many of these so-called respiratory pigments. Of those certainly respiratory, hæmocyanin is by far the most important.

The question as to whether any other pigments besides those of the nature of chlorophyll and the respiratory pigments are of direct physiological importance, is a somewhat difficult one. According to many, some pigments, such as the colouring matter of the Red Algæ and the dark pigments of many animals, are of great importance in protecting delicate tissues from the injurious effects of certain of the rays of white light.

Again, the pigment described by Dr. MacMunn as enterochlorophyll, which is so common in connection with the alimentary tract in many Invertebrates, is supposed by some to be of importance in digestion or assimilation. There are other suggestions of the same kind, but there is as yet little certainty, the functions of the large majority of pigments being quite unknown.

2. Derived Pigments

The second group of pigments includes those which are formed by the decomposition or breaking down of the pigments of the first group. These are probably always useless so far as any direct function is concerned, and it might be thought that they should be simply classed as waste products. It seems, however, better at present to reserve the term waste products, at least in animals, for those results of nitrogenous metabolism to which the term is usually applied by physiologists. We have already mentioned that hæmoglobin undergoes a series of retrogressive changes, resulting in the production of various pigments. Most of these, in Vertebrates at any rate, are of little importance in producing coloration, are speedily eliminated from the body, and in our own species are chiefly of interest to the practical physician. There are, however, some exceptions. Melanin, the dark pigment which colours the hair or skin of most mammals, and which, as we have seen, is indirectly of importance in producing many of the gorgeous tints of birds, is thought by some authorities to be a derivative of hæmoglobin. Again, the frequently exquisite colours of birds' eggs are almost certainly due to pigments derived from the blood. It is quite possible that this group of pigments may ultimately turn out to be a large one, but as yet hæmoglobin is one of the few pigments whose metamorphoses have been fully worked out.

3. WASTE PRODUCTS

The next group, the pigments which are definitely
waste products, or are produced by the modification
of waste products, has only been seriously studied
very recently. The researches of Mr. F. Gowland
Hopkins in this country, and of Dr. Urech abroad,
as well as of others, have demonstrated within the
last few years the extremely interesting and important fact that the colours of the butterflies
belonging to the family of the Pieridæ are due to
pigments which are modifications of the ordinary
waste products of the organism. Hopkins's discovery
that the yellow pigment which he calls lepidotic acid,
found in the wings of the Pieridæ, occurs also as one
of the normal waste products of the organism, is one
of extreme interest to the comparative physiologist.
It is well known that in higher Vertebrates death
follows with extreme rapidity upon the removal of
the kidneys, and it is usually stated by physiologists
that it is the nitrogenous waste products themselves,
or their precursors, which poison the animal ; in the
case of butterflies, however, we have a modification
of uric acid stored up in the wings, functioning as a
colour-producing substance, while the same substance
is eliminated from the body as a waste product at
the time of the emergence from the pupa state. We
are thus forced to suppose that the wings of butterflies, being relatively non-vital parts, can have
poisonous substances stored up in them without
injury to the organism, and that therefore the
utilisation of waste products as colouring agents can

only occur in cases where the coloured structures are not intimately connected with the blood system.

Another substance of this group which among fishes is widely utilised as a colouring agent is guanin. This substance, used in the manufacture of artificial or Roman pearls, is colourless or chalk-like, but it occurs in the skins of fishes as small crystals which frequently display a beautiful iridescence and a pearly lustre. It is to these crystals, mingled with pigments, that the soles, the cod family, and numerous others owe their frequently beautiful colour. The crystals occur in the scales, and also in the deep layers of the skin, in the peritoneum and in the air-bladder, as well as elsewhere. It is interesting to note that in Elasmobranchs, which are ancient fishes, although guanin is present in the skin it has no metallic lustre, such as it exhibits in many of the modern bony fishes. Guanin is not of course a pigment in the strict sense of the word, but it is of much importance in producing coloration in fishes, and its composition makes its occurrence in the skin of great interest. It may seem that undue stress is laid upon these peculiar colouring matters, but they are of much interest, because while of most pigments the chemical relationships are unknown, the substances of the uric acid and related groups have been tolerably well worked out. This is partly on account of their importance in practical medicine, for we all know that the production of an excess of uric acid or its imperfect elimination from the body is in man associated with painful diseases. Now in systematic zoology we distinguish higher from lower animals by the fact that the former, in the current phrase, display more differentiation and more integration. Differentiation displays itself most obviously in an increasing complexity of parts, but integration is shown in the increasingly perfect physiological relation of parts. In the evolution of organisms an increasingly rapid elimination of waste must have been a factor which made for progress ; no machine can work well if it is choked with its own waste. Waste products can therefore be employed as colouring agents only under unusual conditions or in organisms at a relatively low level in evolution. Thus we see that the comparative study of pigments must yield important contributions to comparative physiology in general.

In this connection we may notice the interesting fact that, in discussing the evolution of the fœtal membranes in Vertebrates, Dr. Richard Semon connects the appearance of the allantois (the " hypertrophied urinary bladder ") with the acquisition of the metanephros, a more efficient excretory organ than the simpler mesonephros. It is certain that the blood in the lower Vertebrates contains a far larger amount of nitrogenous waste than that of the higher. We shall see later that as we ascend in the scale in Vertebrates we have a corresponding diminution in the amount of nitrogenous waste products deposited in the skin. Now our acquaintance with the physiology of the Invertebrates is exceedingly limited, and it may well be that a knowledge of their pigments may be of much help in deciding general questions as to the phylogenetic position and the relative importance of structures connected with the excretion of waste.

4. RESERVE PRODUCTS

The next group of pigments includes those which are actually reserve products or are associated with such. Although the literature of the subject contains numerous mention of pigments described as reserve products, it is still doubtful whether the description is accurate in any one of the cases. Carminic acid, the pigment of the coccus insect, is very frequently described as a reserve, but this is doubtful; it is a glucoside, and therefore a carbohydrate is produced by boiling it with dilute acid, but nothing is certainly known as to its function.

The term reserve product is applied with more plausibility to various members of the large series of lipochromes or fat-pigments. The lipochromes are pigments, varying from yellow through orange to red in colour, which in the dry state give a blue colour with concentrated sulphuric or nitric acid, and which are soluble to a greater or less degree in all the solvents of fats, as well as in fats themselves. They are divisible into two series, according as they do or do not form compounds with the caustic alkalies. Both series occur in animals, but the second only in plants. In plants the best known lipochrome is carotin, which is widely distributed, and is, according to Carl Ehring, a cholesterin fat. The chemical nature of the other series is still unknown.

The lipochromes occur frequently, though not invariably, in association with fat in organisms. From the red flesh of the salmon, for example, it is possible to extract an oil containing the pink

colouring-matter in solution. Now during the breeding season, when the salmon is fasting, the fat which
is so abundant in the muscles is transferred from
these to the ovaries, and with the fat the pigment
is also carried, so that the muscles become pale in
colour. A very similar process has been described
by Zopf in the case of a fungus (*Pilobolus*). Here
in the endospores, the gemmæ, and the zygospores,
drops of oil occur intensely coloured with a lipochrome
pigment. When the gemmæ or spores germinate,
the oil-drops disappear, and with them the pigment
also disappears. Zopf in consequence describes this
pigment as a reserve product. It seems, however,
safer as yet merely to admit that lipochrome pigments
frequently occur in association with reserves, leaving
the question as to whether the pigments themselves
are capable of being employed in metabolism to be
determined in the future. It is extremely unlikely
that the lipochromes are always or even usually of
the nature of reserves, as they occur in a variety of
structures where such a significance seems impossible.
They are extremely common pigments both in plants
and animals, and we shall have to recur to them very
frequently in the course of the following pages.

5. Introduced Pigments

The last group of pigments that we shall consider
here includes those which are not produced by the
organism in which they occur, but are obtained from
other organisms used as food and are transferred
apparently almost unaltered to the tissues. The
colouring-matter of green oysters, which was formerly

said to be obtained from the diatoms of their food, used to be given as an example ; another is the case worked out recently by Mr. Poulton (1893), showing that the green pigment of some caterpillars is derived from the green leaves upon which they live. Another example, which in some degree resembles Mr. Poulton's case, is one mentioned by Zopf (1892). In studying the colouring-matter of the fungus *Pilobolus* already mentioned, he found that a parasite growing on the fungus took up not only the drops of oil but also the pigment associated with the oil, the result being that parasite and host were similarly coloured. This is interesting, because it probably has some bearing on certain of the cases of colour-resemblance which are now so numerous in the literature of colour.

The above five sets of pigments have been considered in some detail, because it is quite possible that ultimately the greater number of the pigments at present known may be found to fall into one or other of these categories. In the meantime, however, there is no proof of this, and there are very many pigments which it is quite impossible to classify. On this account, we shall simply consider them systematically under the organisms in which they occur. Before doing this it may be well, in the absence of a complete classification of *pigments*, to consider for a little the distribution of *pigmental colours* in organisms.

THE DISTRIBUTION OF PIGMENTAL COLOURS

Blue as a pigmental colour is exceedingly rare among the more complex animals, but is not uncommon among plants and the simpler forms of animal life. Thus the blue colours of many butterflies, of the feathers and skin of birds, of the mandrill among mammals, are all due to structural coloration, while the blue of hyacinths, of some jelly-fish, and of the lobster are pigmental colours. In flowering plants, as will be seen in the chapter on plant pigments, a blue colour is in some cases associated with an alkaline condition of the cell-sap, but beyond this there is still much uncertainty. In animals a blue colour occurs especially in surface forms, whatever their relations, *e.g.* jelly-fish, molluscs as in *Ianthina*, and Tunicates. It is also remarkable as one of the conspicuous colours of coral-reefs where it occurs in corals, sea-anemones, Turbellarian worms and starfishes ; the blue colours of the cuttles and fishes of the same situation are no doubt optical. In general, therefore, in animals blue pigmental colours occur in organisms of simple structure exposed to strong light. A comparative investigation of the blue pigments of surface animals would be of great interest, but has yet to be made. The two main hypotheses as to origin are on the one side that of the protective value of the colour, and on the other of the direct action of light as the important factor ; it is not improbable that it may turn out that it is abundance of free oxygen which is the really important point. In general the blue pigments of animals are characterised by their usual solubility in water, their instability, and the fact that they frequently give either the litmus reaction or some modification of it on acidification. To what extent this indicates affinity among them is at present quite unknown ; apparently most are readily reduced.

Looking at the distribution of blue pigments more in detail, we find that while blue and purple colours are common in the Flowering Plants, there is much uniformity in the pigments, which seem all to belong to the *anthocyan* series. In the lower plants blue pigments are more numerous. Thus various blue and purple pigments have been described among Bacteria—the blue or purple colour which develops on decaying meat is a good example. These pigments have been compared by some to natural aniline dyes. With these, and as yet of little interest except to the chemist, we may put the bluish-green colours of certain Fungi.

A more interesting pigment is phycocyan, which occurs in the bluish-green Algæ (Cyanophyceæ). It is a beautiful pigment, blue in transmitted and blood-red in reflected light, and is said by Molisch to be of albuminoid nature. It is soluble in water, and has been obtained in crystalline condition.

In animals blue pigments seem to be more numerous than in plants. They are found in various Protozoa, as in *Nassula* and *Stentor*; in the former it seems possible that the pigment is taken up from the blue-green Algæ of the food. In the Cœlentera blue pigments are exceedingly common, especially in shallow-water forms in warm seas, and in the pelagic jelly-fish. In worms and in Echinoderms we find that, although blue colours occur, the pigments producing them do not appear to have been isolated. Similarly in insects, where blue is very common, blue pigments have not yet been described, although it is improbable that the blue in all cases is structural.

In Crustacea there is probably only one series of blue colouring-matters, the so-called soluble blues, which will be more fully described in the chapter on the group.

Among Mollusca we have notably the blue pigment of *Ianthina* and the purple one of *Aplysia*, the undescribed colouring matters of the shells of mussels, of the mantle of *Tridacna*, and of others.

Among Vertebrates blue pigments do not appear to have been as yet described, except in the case of the blue pigments of the eggs of many birds. These pigments are apparently derivatives of hæmoglobin.

A *green* colour is of course almost universally distributed among plants, where it is due to the important pigment chlorophyll. Among animals green like blue is rare as a pigmental colour, except in simple organisms. The questions connected with the green colours of animals have been greatly complicated by the common habit of hastily assuming that any pure green colour in animals is likely to be due to chlorophyll. Chlorophyll occurs apparently in many Protozoa, in *Hydra viridis*, in the fresh-water sponge *Spongilla*, and in some Turbellarian worms. Its occurrence in numerous other animals, especially worms and insects, has been often asserted, probably in some cases because the gut contained unaltered chlorophyll which gave the characteristic spectrum. "Modified chlorophyll" is said by Mr. Poulton to colour the tissues of many caterpillars. A group of pigments which resemble chlorophyll, and which have been often mistaken for it, are apparently widely spread among Invertebrates. They usually occur in connection with internal organs, and are therefore only in rare cases of direct importance in coloration. Of this group, "enterochlorophyll," bonellin, and chætopterin are good examples ; the two latter give rise to external coloration in the forms in which they occur. We shall make frequent reference to these pigments in the course of the following chapters.

The occurrence of green colour in organisms from which yellow lipochromes only can be extracted is an exceedingly common phenomenon, which has been observed in many different animals. The green colour is doubtless due in many cases to the structure of the coloured parts ; in others it may be due to a combination of the lipochrome with a base ; there is little doubt that the green colour of many Crustacea is in part due to this cause. Apart from chlorophyll and lipochrome combinations, green pigments occur freely in many Cœlentera, in many worms, and in some Mollusca; in these cases they are usually soluble in alcohol and often give banded spectra, but most have been very imperfectly investigated. In Vertebrates green pigments are almost absent ; as in the case of blue pigments, the most marked exception is probably the green pigment which colours birds' eggs, though according to Sorby the green colour is here produced by a mixture of blue and yellow. As a colour, green in animals is most abundant among those inhabiting trees or herbage, but is here very frequently structural. In marine animals it seems to occur in large masses chiefly in coral-reefs, where the corals are often largely green. In these situations green plants are relatively rare, and accordingly the green colouring-matter has been supposed by Dr.

Hickson to perform the same functions as chlorophyll. It is apparently not chlorophyll (see Chap. IV.).

Yellow is an exceedingly common and widely distributed colour, occurring most frequently, however, in animals in which other pigments are also present. It not infrequently occurs in the form of spots or stripes upon a dark ground, although it is common in flowers as a ground colour. In most cases yellow is due to the presence of yellow lipochromes, which are perhaps the most universally distributed of all pigments. In view of their exceedingly wide distribution, it is perhaps remarkable that yellow animals should not be more common than they are. Apart from the yellow lipochromes we have as important yellow pigments lepidotic acid, the waste product which occurs in the wings of some butterflies, and the yellow pigments of the eggs of birds. The yellow or tawny colour of the hair in many mammals is not due to a special yellow pigment, but to the uniform distribution of a small amount of dark pigment.

A *red* colour is perhaps always due to pigment, and red pigments are fairly numerous. Large numbers of red animals are coloured with lipochrome pigment; the red lipochromes indeed begin at the Protozoa and extend upwards to birds. Their colour varies from deep orange to pure rose-red, but all have an indescribable fatty appearance which makes them readily recognised by any one accustomed to working with lipochromes. Red is very common among marine animals, especially among Crustacea, Echinoderma, some Mollusca, Protozoa, and Cœlentera. In Crustacea it occurs among some surface forms, but attains its greatest brilliancy among those inhabiting deep water, which often exhibit a wonderfully pure and bright scarlet tint. All these red tints are due to lipochromes.   The red blood-pigment hæmoglobin is frequently important in coloration in simple forms, especially worms, but the effect is produced by the shining through of the red blood, and not by the deposition of hæmoglobin as a superficial pigment.   In butterflies red pigments important in coloration occur which are probably modified waste products.   Among plants red pigments are probably usually either lipochromes or anthocyans ; red is an exceedingly common colour among plants.

Black and deep brown pigments are very widely distributed in the more complex animals, and are virtually absent from plants and · simple animals. As exceptions we may note the presence of a brown pigment in *Hydra fusca* and the " brown body " of Polyzoa.   The former, according to Miss Greenwood, is a waste substance, the latter will be discussed under Polyzoa.   That their distribution is not wholly dependent upon structure is, however, shown by the fact that they are rare in Crustacea and exceedingly common in Insecta, slightly developed in most Chætopoda and often marked in Hirudinea.   There is some evidence to show that they do not commonly occur in connection with structures containing carbonate of lime, which may serve to explain their rarity in Crustacea.   At least it is noticeable that in Mollusca very dark brown or black pigments do not usually occur in the shells, while they may be common in the mantle, internal organs, or secretions, as in

cuttles. The question whether there is any genetic
connection between the dark pigments of animals
belonging to different orders is one of great interest,
but one which it is at present impossible to answer.
Dark pigments are mostly very stable and insoluble;
where they have been analysed they have been
usually found to contain nitrogen, and in some cases
sulphur as well. They are probably of no further
use in metabolism, and seem often to tend to increase
in the course of development, and to be more abundant
in dominant than in weak species. Many have
regarded them as directly waste products increasing
with increase of metabolism, a question which we
shall discuss later.

Pale brown pigments may arise in so many
different ways that nothing of a general nature can
be profitably said of them. In plants and perhaps
in some insects tannin may play the part of a brown
pigment.

White as a pigmental colour is rare. Often
purely structural, it is sometimes due to the deposition of fat in the subcutaneous tissues. In the
Pieridæ among butterflies white is due to uric acid
which here plays the part of a white pigment.

THE SPECTROSCOPIC CHARACTERS OF PIGMENTS

The pigments with which we have been concerned
in this chapter are recognised by their colour, their
reaction to various chemical reagents, and finally by
their spectroscopic characters. Most people are
familiar, at least by hearsay, with the important part
which has been played by the spectroscope in

modern chemistry, and have heard of the wonderful discoveries as to the composition of the sun and some of the stars which have followed from the spectroscopic examination of the light emitted by these bodies.   Now we have already seen that many of the pigments of plants and animals are extremely fugitive and unstable, many of them also are exceedingly difficult to extract from the tissues. By means of the spectroscope, however, the coloured tissues may often themselves be directly examined, and on account of the readiness with which the operation may be performed, a large number of pigments are known chiefly by their spectra.   Now even in the case of the metals, we find an authority like Mr. Crookes saying that "inferences drawn from spectrum analysis *per se* are liable to grave doubt, unless at every step the spectroscopist goes hand in hand with the chemist."   If this be true of radiant matter spectroscopy, where the spectra are capable of extraordinarily accurate study, how much more likely is it to be applicable to the spectrum analysis of pigments, where the methods are as yet clumsy and inadequate!   Thus we find as a matter of practical experience, that it has been found impossible in the aniline industry to employ the spectra as a test for affinity.   Again, in spite of the common assertion that pigments yielding identical spectra are themselves identical, we find that three distinct pigments—hæmoglobin, carmine, and turacin —are described by competent authorities as giving spectra which are virtually identical.   Turacin, a pigment discovered by Professor Church, is of purplish-violet colour, and occurs in the feathers of

certain birds.   It is incapable of existing in states
of oxidation and reduction, and contains copper and
not iron, so that it is unlikely that it is closely
related to hæmoglobin.   Carmine again is not
nearly related either to turacin or hæmoglobin.
The spectrum test in this case seems to break down
utterly.   In spite of numerous difficulties of this
kind, the literature contains numerous identifications
of pigments based only upon the use of the spectroscope ; thus a recent observer describes chlorophyll
in the skin of the horse !   In the following pages
identifications based only upon spectra have been
almost entirely omitted, for the evidence is considered insufficient.

CHAPTER III

THE COLOURS AND PIGMENTS OF PLANTS

Pigments and Colours of Bacteria and Fungi—Chlorophyll
and the Associated Pigments—Colour and Pigments of
Algæ—Pigments of Flowering Plants—Autumnal Coloration—Colours of Flowers and Fruits—Meaning of Plant
Pigments and Summary.

In considering the colours and pigments of plants
we may conveniently begin with the colours of
Bacteria and of Fungi, and then pass to the consideration of the colouring of chlorophyll-containing
plants.

The colours of Bacteria are often surprisingly
brilliant. The red spots which *Micrococcus prodigiosus*
forms on moist bread, the violet colour which
sometimes appears on decaying meat, are familiar
cases in point, but less familiar forms are often
equally bright. There are some facts of interest
in regard to the position in which the pigments
occur, and the conditions necessary for their formation.

As to the first point, the pigment may occur
within the cells of the colony. In 1873 Professor
E. Ray Lankester described an interesting peach-

coloured Bacterium found in stagnant river-water, in which the peculiar pigment was confined to the protoplasm of the individuals and did not extend into the jelly surrounding the colonies. Mr. Slater has similarly described a form producing a bright red pigment which is confined to the cells. On the other hand, in the case of *Micrococcus prodigiosus*, the red pigment is confined to the masses of mucilage surrounding the colonies and does not occur within the cells at all, while in yet other cases the pigment may be found only in the substance in which the colonies are living.

As to the conditions under which the pigments are produced, it is well known that *Micrococcus prodigiosus* loses its power of producing pigment at high temperatures, while light is necessary for the production of the purple pigment of *Beggiatoa roscopersicina*. This pigment is of especial interest because, according to Engelmann (1888), its presence in the organism is associated with the power of breaking up carbonic acid and setting free oxygen, therefore it has a function equivalent to that of the chlorophyll of green plants. In the case of *Bacillus pyocyaneus*, according to Gerrard, the power of pigment production is dependent upon the nature of the medium, as well as upon the particular race of the microbe.

As a point of interest with regard to the chemical nature of the pigments, we may notice that in 1889 Zopf described a yellow pigment formed by *Bacterium egregium* as a lipochrome, and stated that this was the first time that lipochromes had been described as products of bacterial action; he has since described other cases. The bright red *Micrococcus* pigment is

soluble in alcohol but not in water, and is said to be
not improbably related to the aniline dyes—a statement which has also been made for the blue pigment
found on decaying meat.   The pigment of the peach-
coloured Bacterium described by Professor Lankester
was found to be very insoluble, but gave a distinct
three-banded spectrum.

The pigments of Bacteria have been chiefly studied
as a means of identifying the organisms producing
them, and a further account of them here is unnecessary.   Their main interest lies in their frequent
brilliancy and variety.   Bacteria are remarkable for
the enormous number of chemical substances which
are the result of their activity ; it would be strange if
among these there were not some capable of producing colour, but the remarkable brightness of many
of the tints is of considerable theoretical importance.
The fact that the production of the pigment is a
factor both of external conditions and of the constitution of the organism suggests that bacteriologists
may one day have something to say on the general
problem of the conditions necessary for pigment
formation.

With regard to the higher Fungi, it is a familiar
fact that they frequently display great brilliancy of
colour.   This is especially true of their fructifications
which on barren ground are often a not inconsider-
able factor in the production of the bright tints of
autumn.   Although these " toadstools " are regarded
with disfavour by many people on account of their
uselessness for culinary purposes, their beauty,
especially when seen in natural conditions among
yellow bracken and glowing brambles, can hardly be

denied. It is not, however, the conspicuous forms only which are brightly coloured. A species of *Peziza* produces the brilliant green colouring matter which so often stains decaying wood, and the yellow spots which *Æcidium* causes on the leaves of the barberry are common enough objects in spring. Zopf has devoted considerable attention to the pigments of Fungi, and has examined many of them. He has shown that many Fungi contain one or several different lipochromes, and, as has been already mentioned, these lipochromes may be taken up unaltered by parasites. We have also already noticed Zopf's view that the lipochromes function as reserves. Besides the lipochromes, a very large number of pigments in Fungi have been described and named. The chemical characters of some are known, of others unknown, but in few cases, if any, has any definite function been assigned to them. Thus we find in the Fungi a great number of pigments of very diverse colours and chemical composition and of unknown function. Notwithstanding this last fact, the coloration exhibits some at least of the characters displayed by organisms whose colour is thought to be of supreme importance to them. In the large toadstools the bright colour is usually confined to the upper and conspicuous surface of the *pileus*, and this surface is not infrequently marked or spotted with great regularity. These facts are of considerable importance with regard to theories as to the (proximate) origin and meaning of colour. They show us that not only pigments, but also regularity of coloration may occur in organisms in cases where the relations to other organisms are too simple for us to suppose that either pigment or coloration can be of direct use. When we come to consider the phenomena of colour in organisms whose relations to other organisms are extremely complex, we shall find that there is a constant tendency to look for the cause or the justification of the colour phenomena in these complex relations. It is therefore important to realise that it is illogical to seek for a complete explanation of colour phenomena in complex organisms while those of relatively simple ones remain unexplained.

CHLOROPHYLL AND THE ASSOCIATED PIGMENTS

This brief account of the colours of the plants not containing chlorophyll must suffice, for it is impossible here to go into further details. We shall treat the colours and colouring-matters of the chlorophyll-containing plants in greater detail, for in them the colours are as a whole more striking and beautiful, and the pigments are much more fully known.

In these plants chlorophyll is of course by far the most important pigment, and sometimes indeed the only one. For details as to its chemical characters reference must be made to the text-books, but we may here recall one or two points as to its occurrence. Chlorophyll occurs, generally speaking, in all plants except the Fungi ; it is especially abundant in the leaves or vegetative organs, and occurs in association with definite parts of the cell protoplasm—the chlorophyll corpuscles. It is frequently associated with other pigments which may mask it, or may replace it in special regions of the plant. This partial replacement is especially noticeable in the

Flowering Plants, and it is only here that it is seriously regarded as the result of a process of selection. The unequal distribution of chlorophyll is, however, well marked in some of the lower plants.

As to the associated pigments, we find that in Flowering Plants the chlorophyll of green leaves is always mingled with a greater or less amount of a yellow lipochrome usually known as xanthophyll. The different tints of leaves are caused in great part by the varying amounts of the two pigments present, a large amount of xanthophyll giving the leaf a yellowish tint. The exact relation of this yellow pigment to the colouring-matter of leaves blanched by growing in darkness is uncertain, but the two are probably at least nearly related. We know from investigations on the lipochromes of animals that these pigments are not only peculiarly unstable, but that a number of them frequently occur simultaneously in an organism, and are then very difficult to distinguish from one another. This fact enables us to understand readily how it is that the questions whether there is more than one lipochrome associated with chlorophyll, and whether etiolin (the colouring-matter of blanched leaves) is or is not identical with xanthophyll, remain undecided. In considering the colours of flowers and fruits we shall find that xanthophyll, or pigments closely related to it, may play an important part in the economy of the plant, but the function of the xanthophyll connected with chlorophyll is unknown, if indeed it possess any. The only suggestion that has been made in this matter is that the xanthophyll may have some function in protecting the protoplasm of the cells from the injurious effect of certain of the elements of white light by absorbing them, or that it in some way assists the assimilating action of chlorophyll. As yet, however, there is little evidence for either of these suggestions, and the majority of authors are quite silent as to the function of xanthophyll.

COLOUR IN ALGÆ

A similar association of chlorophyll with other pigments, especially lipochromes, is often well seen in the Algæ, where the pigments seem more varied than in flowering plants. In many cases they completely mask the chlorophyll, while in others the unequal distribution of the chlorophyll produces colour effects which show a striking resemblance to those seen in the flowering plants.

The masking is especially well seen in the case of the colouring-matter of the Florideæ or Red Seaweeds, which has further had an important function ascribed to it. These Algæ for the most part live in deep water, and are chiefly known to those who are not botanists by the beautiful reddish-pink fronds of *Delesseria* so often cast up on the seashore during the summer months. Like *Delesseria* the Florideæ are nearly all red or violet when living, but if placed in cold, fresh water the red pigment dissolves out and leaves the seaweed green. The solution in water of the red pigment, known as phycoerythrin, fluoresces strongly from red to yellowish-green. The chlorophyll with which it is associated in the Florideæ is said to differ in some respects from the chlorophyll of other plants. Now, as we have said, the Red

Algæ are typically deep-water forms, and it is well known that water absorbs the various rays of light unequally, so that the yellow and red rays do not penetrate to great depths, but are stopped long before the blue and violet. Engelmann (1883) experimented upon the effect of the different rays of the spectrum upon the process of assimilation in various plants, and found that while green plants, generally speaking, assimilate best in red light, the Red Algæ assimilate best in blue light, and can therefore live at depths impossible to other plants. This does not of course assist us as to the meaning of the phycoerythrin, but Kerner has supplemented Engelmann's observations by the suggestion that the red pigment absorbs the blue and violet rays and converts them into red rays, so that according to him the presence of the red pigment is absolutely necessary to these Algæ. The suggestion is perhaps chiefly of interest because it is one of a great number which have been made lately as to the effect of pigments on the incident rays of light, and which assign a physiological importance to pigments hitherto regarded as useless. Nothing is known as to the chemical affinities of phycoerythrin. It is possibly of proteid nature, and like numerous other pigments has been, on extremely doubtful evidence, accredited with a respiratory function.

We do not propose to go into further detail on the subject of pigments associated with chlorophyll, but may mention one or two cases in the Algæ of that replacement of chlorophyll by other pigments in special regions which is so common in the higher plants. In many of the unicellular Algæ, as in many of the Protozoa, there is a constant oscillation between red and green as a ground colour. It seems most probable that this, as in higher plants, is due to a destruction of the chlorophyll green and a consequent predominance of an associated lipochrome. In Algæ it occurs nowhere in such an instructive way as in the brittle-worts *Chara* and *Nitella*. Here virtually the whole plant is coloured green by chlorophyll, but the reproductive organs, and especially the antheridia, as they ripen become a bright red, the colour being due to the red chromatophores which replace the chlorophyll corpuscles. The whole process is exactly analogous to that which occurs during the ripening of red fruits like the rose-hip. Quite similar is the process which occurs during the maturation of the antheridia of brown seaweeds like *Fucus*. Here the plant owes its colour to a combination of chlorophyll green, a brown pigment, and a lipochrome. In the oospheres all these three are retained, but in the antherozooids the brown and the green disappear, and the orange-coloured lipochrome remains in the chromatophores and gives rise to the orange coloration. The retention of the chlorophyll in the female elements and its disappearance from the male many would regard as an illustration of the greater vegetativeness of the female. These two examples may suffice to show that the processes which give rise to the colours of Angiosperms have very well-marked analogues among Cryptogams.

PIGMENTS OF FLOWERING PLANTS

Among the flowering plants chlorophyll is of course the supremely important pigment. With it as already seen are associated lipochrome pigments whose nature and amount determine the exact shade of green displayed by the vegetative organs. In flowering plants not only are the organs connected with reproduction often brightly coloured, but the vegetative organs themselves may also display brilliant pigments. Among the pigments two series are of special importance—these are first the lipo-chromes, and second the anthocyans. Of the lipo-chromes it is not necessary to say anything further at present, but the anthocyans merit more detailed consideration.

Anthocyan, or the series of pigments included under this name, occurs dissolved in the cell-sap, and varies in colour from blue to red. It is an exceedingly common pigment in the higher plants, occurring alike in vegetative and reproductive organs, and is readily soluble in water. By steeping the rind of an apple or slices of beetroot in water, a red solution of the pigment is readily obtained. If an alkali such as caustic soda or ammonia be cautiously added, the colour changes from red to blue, green, and yellow successively. Finally, on adding excess of alkali the solution becomes colourless. If the alkali added be ammonia, this may be removed by boiling, when the blue colour will once more reappear. This change from red to blue is of course the litmus reaction, so familiar to all who have worked in a chemical laboratory. It is exceedingly characteristic of certain of the anthocyans, and is of much importance in the production of the colours of flowers and fruits.

The chemistry of the anthocyans has not yet been fully worked out, but it seems probable that they are derivatives of tannin, probably oxidation products. The colouring-matters of the different varieties of grapes, as well as of the leaves of the vines bearing them, are said to be due to oxidation products of the tannic acids which are so abundant in these plants, and these colouring-matters are undoubtedly of anthocyan nature. There is also other evidence pointing in the same direction. Tannin is probably a substance of no further use to the plant, and therefore it is probable that the anthocyan pigments come into the group of waste products. We say *probable* only because, while plant physiology is so imperfectly known as at present, it seems hardly justifiable to apply to plants terms which in relation to animals have a perfectly definite meaning.

As to the proximate importance of anthocyan to the plant, there are endless suggestions and hypotheses. Into all these we cannot enter here, but as a type may take a recent interesting paper by Professor Stahl on the meaning of anthocyan in brightly coloured foliage leaves.

Stahl studied the bright colours of the leaves of certain tropical plants, and decided that these could not be regarded as warning colours—a conclusion which is probably well founded! He therefore set himself to discover some physiological function. He found by experiment that the red parts of plants absorbed 1.82 per cent more heat than green parts, and he considers that this fact explains the peculiar distribution of this red colouring-matter, its presence in young shoots, Alpine plants, etc. He believes that it also explains its occurrence in wind-fertilised flowers, in many Dicotyledons, in Gymnosperms, in Cryptogams, and so on, for the increased absorption of heat in these cases would aid the growth of the pollen tube, etc. It does not, however, explain the occurrence of red pigment in the leaves of tropical plants, where it is often associated with the simultaneous occurrence of white spots, which have, of course, exactly the contrary effect as regards heat absorption. Stahl finds that such plants inhabit shady places, and his opinion is that the simultaneous existence of red and white spots aids transpiration by producing an unequal absorption of heat. The white spots are produced by the presence of large air-spaces beneath the epidermis, the result being that the spots cool more slowly at nights and so receive less dew, and therefore render transpiration possible even at a low temperature. The paper contains some other observations which we need hardly note here. The reviewer of Stahl's paper in the *Botanische Zeitung* calls it a model of biological investigation, and in its lavish employment of ingenious hypotheses it may at least be regarded as a type of many biological investigations.

Another investigation which seems to throw more light upon the peculiarities of the distribution of anthocyan is one by Molisch. In working at anthocyan, Molisch (1889) attempted to obtain a solution from the brightly coloured leaves of some species of *Coleus*. He found, however, that on boiling these leaves with water, he did not obtain a coloured solution, although the leaves themselves lost their colour. If, however, the colourless solution be evaporated to dryness, anthocyan is left behind in its blue or violet form. Similarly if the decolorised leaves be dried at a gentle heat they regain their bright colours. The reason for this remarkable phenomenon Molisch found to be as follows. During life the cell-sap in these coloured leaves is faintly acid or neutral in reaction, but the protoplasm is as usual distinctly alkaline. At the moment of death the alkaline protoplasm mingles with the cell-sap, and the alkalinity of the protoplasm is strong enough to completely destroy the anthocyan colour (cf. action of alkalies as explained above). Molisch also found that leaves which are not decolorised on boiling are remarkable for the large amount of acid contained in the cell-sap, this being apparently large enough to neutralise the alkalinity of the protoplasm. By further experimentation Molisch convinced himself that the decolorisation only occurred when the cells containing anthocyan were in contact with cells very rich in chlorophyll. This he illustrated by a very pretty experiment. He took the leaves of a species of Saxifrage, *Saxifraga sarmentosa*, in which the epidermal cells of the leaf are very rich in anthocyan, and removed a small part of the epidermis from a leaf. Both the piece of epidermis and the remainder of the leaf were then kept at a temperature sufficient to kill the cells, and so bring the cell-sap into contact with the protoplasm. In a quarter of an hour the part of the epidermis still in contact with the leaf became decolorised, while the severed portion completely retained its colour. As the chlorophyll can hardly be supposed to have a direct effect on the anthocyan, Molisch concludes that the " conditions for the formation of alkaline substances must be especially favourable in chlorophyll-containing cells." This is a conclusion of great interest and importance. If cells which contain much chlorophyll also contain strongly alkaline substances, and therefore, as we may reasonably suppose, tend to destroy any anthocyan pigment which may be formed, then this is equivalent to saying that the conditions which are favourable to the development of chlorophyll are unfavourable to the development of anthocyan, and *vice versa.* This is a very tempting conclusion, for it seems to explain many points in the distribution of anthocyan which have hitherto been very puzzling. Broadly speaking, anthocyan tends to appear conspicuously in many leaves and shoots in early spring, in many leaves in autumn, in flowers, and in fruits. Its function is often said in the case of vegetative organs to be the protection of chlorophyll from the injurious effects of excessive light, but this does not account for its appearance in autumnal leaves. If, however, we may assume that chlorophyll and anthocyan are in antithesis to one another, we can readily understand why the latter should appear in spring, before the power of assimilation is completely established, in autumn when it is beginning to disappear, in ripening flowers and fruits where it is more or less completely lost, and also in the leaves examined by Stahl which grew in shady places, in conditions unfavourable to the development of much chlorophyll.

AUTUMNAL COLORATION

From this consideration of the common *pigments* of the higher plants, we may pass to an account of the peculiarities of the *colours*, selecting as examples the colours of autumnal leaves and of flowers and fruits.

We are all in this climate familiar with the fact that the chlorophyll of stems and leaves is short-lived : we all know how the delicate yellow-green of spring leaves deepens into the dark green of summer, and then disappears in the yellows and reds of autumn, while these in their turn lose their glory before the chill blasts of winter. Chlorophyll is bound up with the assimilating power of the plant, and as this power diminishes, the chlorophyll which is its outward expression disappears also. This is, however, true only of the green colouring-matter, the chlorophyll-green, and not of the associated pigments. The chlorophyll-green is probably reabsorbed along with starch and any other useful substances which may be in the leaf, while the yellow xanthophyll remains behind in the form of oily drops, set free by the disintegration of the chlorophyll corpuscles. In the simplest case, *e.g.* that of straw, there is thus produced a uniform yellow coloration, the chlorophyll being completely removed and the xanthophyll only left. To produce the splendour of our October woods, however, other factors have to be introduced. In the first place the removal of the chlorophyll is often

F

slow and partial, giving rise to the appearance of a variegated leaf. Further, owing either to the formation of several lipochromes, or to the unequal distribution of a single lipochrome, parts of the leaf become a deeper shade of yellow or orange, sometimes becoming intensified to a dull red, while in some cases there is produced a special red anthocyan pigment. There are therefore three main factors in the production of the tints of autumn: (1) the disappearance of the chlorophyll-green, (2) the increasing prominence of the lipochromes, and (3) the development of anthocyan. Other changes of minor importance also occur. Thus the general effect is often heightened by the dull brown colours assumed by the leaves of such trees as the oak and the beech. These colours are produced by the oxidation of the tannins of which these trees contain such an abundant supply. These substances are probably useless, and are got rid of in the falling leaves and the bark. Although these changes tend to occur with great regularity every autumn, it is a matter of common experience that they are to a large degree dependent upon the weather, a fine dry autumn with a touch of frost being specially favourable to the development of brilliant colouring. Autumn colouring is of great interest in a comparative study of coloration. There is no reason to suppose that the colouring is of the slightest use to the trees, and yet it often displays to an extraordinary degree that beauty and perfectness which we are accustomed to regard as the result of the action of Natural Selection. It is further of fundamental importance in the investigation of the

causes of the colours of flowers, for the three
processes already mentioned as the factors in the
production of autumnal coloration are precisely the
same as the processes which produce the colours of
flowers.   If the tints of autumn arise naturally from
the check to vegetation produced by the first breath
of frost, then we may reasonably suppose also that
the colours of flowers are also in origin the natural
result of diminished vegetative power.

Although in our climate the majority of our
plants shed their leaves in autumn in a cloud of
glory, yet we have of course some which retain their
leaves for several years, or do not shed them all
at once, being, as we say, evergreen.   Even here,
however, the diminished vegetative power is frequently
seen in the partial modification of the chlorophyll.
The leaves of many evergreens assume a reddish
colour in winter, and this is due to the partial disappearance of chlorophyll from the corpuscles, and
its replacement by red oily drops, probably of
lipochrome nature.   These red drops disappear again
in spring when the leaves assume their normal green
colour.   This red pigment differs from that of most
autumnal leaves in being confined to the chlorophyll
corpuscles, while in most cases a red colour is due to
anthocyan dissolved in the cell-sap.

COLOURS OF FLOWERS AND FRUITS

The subject of autumnal coloration leads up to
the colours of flowers and fruits.   These in the
general case are due either to anthocyan pigments
dissolved in the cell-sap, or to lipochrome pigments

contained in solid bodies, known as chromoplasts or chromoleucites.

Anthocyan pigment colours the petals of hyacinths, bluebells, roses, etc., and such fruits as grapes, blaeberries, cranberries, and so on. The property of colour-change which we have already seen it to possess is of considerable importance in the production of the colours of flowers and fruits, for when anthocyan is present in, for example, the cells of petals, its tint depends in part on the degree of acidity of the cell-sap.

Thus we are all familiar with the change in the colour of the flowers as they develop, which is frequently so conspicuous a feature in various members of the natural order Boraginaceæ. The forget-me-not is pink in bud and blue when full-grown, the pink colour occasionally persisting as a variation. The colour-change is associated with a diminished acidity of the cell-sap of the cells of the petals. The sap is at first strongly acid, but as the flower develops the acid disappears. Most flowers which in natural conditions are blue show as a variation, or under cultivation, a tendency to become pink, e.g. pink hyacinths, pink Campanulas, etc.; a fact which seems to indicate that the amount of acid present in plants tends to vary, or is in an unstable condition. Such a variation though most common in cultivation probably also occurs in natural conditions. Thus the common milkwort, Polygala vulgaris, may be found in the same locality under at least three different varieties, with blue, pink, or white flowers respectively. It is reasonable to suppose that this variation is the result of variation in the amount of acid present in the cell-sap.   Other plants again show an extraordinary constancy in the tint of the anthocyan of the petals, the fruitlessness of florists' efforts to produce blue roses is said to be the result of some peculiarity of constitution of the pigment which makes the tint impossible.

The colours of the lipochromes vary from yellow through orange to red, and they colour such flowers as daffodils, jonquils, yellow and red lilies, such fruits as those of the tomato, the melon, the honeysuckle, the asparagus, the lily-of-the-valley, and so on.   It is interesting to note, however, that the pure red lipochromes, as distinct from the orange, are rare if not entirely absent in plants (see Chap. II. under Lipochromes).   As to origin, the chromoplasts arise in much the same way as chlorophyll corpuscles do, and indeed in many cases the unripe fruit or undeveloped floral leaves contain ordinary chlorophyll corpuscles.   As development proceeds the chlorophyll disappears just as in the case of autumnal leaves, and the highly coloured lipochrome is left.   Various names have been given to the different lipochromes of flowers and fruits, but there is some reason to suppose that they all arise from the pigments normally associated with chlorophyll.   Sometimes the lipochromes occur in ripe fruits and flowers as crystals, the original envelope of protoplasm having entirely disappeared.   The colours of most flowers and fruits are caused either by anthocyan or lipochrome pigments, or by a combination of the two, but there are a few which are coloured by a yellow pigment of unknown affinities dissolved in the cell-sap.   Such are the colouring-matters of the orange and of the dahlia.

We have thus considered the common colouring-matters of flowers and fruits, and seen that in general terms they fall into two groups, the fixed or lipo-chrome pigments, occurring in the form of solid particles in the cell, and the free or anthocyan pigments occurring in solution in the cell-sap, and varying in tint according to its reaction.   In flowers, further, we frequently find a pure white colour, which we have already seen to be an optical colour caused by the presence of air-spaces between colourless cells. Colours which are not primary colours are caused by a superposition of differently coloured elements, or of colourless elements on coloured ones.

Besides the actual bright colour, flowers are often conspicuous by the beauty and regularity of their markings.   Sometimes this is of exceedingly simple nature, and bears an obvious relation to the nature of the parts.   Thus in the snowdrop the petals are delicately veined with green, and in the wood-sorrel with violet.   A veining with green recalls the appearance of some autumnal leaves when the chlorophyll seems to linger longest at the sides of the veins of the leaf.   In the same way the purple veins of wood-sorrel (*Oxalis*) are quite similar to the reddened veins of many leaves in spring, and just as in leaves the red colour may occasionally spread over the whole leaf, so it is not very uncommon to find in the wood-sorrel as a variation that the petals have become completely purple.   This occurs especially in shady places.   The markings of many petals are, however, of far more complex nature, and bear no obvious relation to the structure of the parts.   Such are, for example, the markings on the labellum in many species of orchids, and on the corolla of the foxglove. There seems little doubt that these markings are in many cases employed by insects as landmarks in their search for honey, and they have been in consequence termed honey-guides. In many cases, however, they are much more complex than seems necessary for this function, and are by no means limited to flowers containing honey ; their meaning and origin are still very doubtful.

Meaning of Plant Pigments and Summary

Looking at the colours of flowers and fruits as a whole, we may say that all the processes which give rise to their brilliant colours have a parallel in the vegetative shoot. The prominence of lipochromes and the development of anthocyan are paralleled in the autumnal coloration of leaves. The development of a white colour is paralleled by the occasional partial albinism of leaves, which occurs either as a result of injury by other organisms, or in some instances as a natural condition, *e.g.* in the lungwort (*Pulmonaria*). Even the complex markings of petals are dimly foreshadowed in the veining of leaves.

All this is fairly obvious, but when we attempt to discuss further the prime meaning of colour in plants, the difficulties are very great. Of the five groups of pigments described in the last chapter, we have in plants the first represented by chlorophyll ; the third possibly represented by the anthocyan pigments, which are apparently derived from tannins, and are probably useless substances ; the fourth, according to Zopf, represented by the ubiquitous lipochromes; and even the fifth group is mentioned by Zopf as being represented among the Fungi. It is curious that in spite of the fact that chlorophyll is such a complex pigment, and can be made to yield numerous coloured derivatives in the laboratory, there is no evidence that in natural conditions any of the plant pigments are produced by its decomposition. The assertion that the xanthophyll of leaves is derived from it has been repeatedly made, but not on any good grounds.

As to the primary meaning of chlorophyll in metabolism we know nothing. Macallum seems to believe that it is in some way directly connected with the nuclear chromatin.

As to the lipochromes, we know neither their primary meaning nor their proximate use. They are perhaps universally distributed in plants, and occur in association with reserves in the shape of oil and fat. They are probably always either yellow or orange, the pure red lipochromes not occurring in plants, and these yellow lipochromes are perhaps invariably found in association with fat. We have already seen that Ehring regards the carotin of the tomato as a cholesterin fat.

The remaining pigments are probably all of the nature of waste or useless substances. Among them are included the anthocyans, the colouring-matters of bark, and of some woods like brasilin, hæmatoxylin, etc. The pigments of Fungi are more numerous than those of flowering plants and are little known.

CHAPTER IV

THE COLOURS OF PROTOZOA, SPONGES, AND CŒLENTERA

Pigments of the Protozoa, their Characters, Variations, and
probable Origin—Coloration of Sponges and Cœlentera
—Distribution of Colours—The Colours of Corals and
Sea-Anemones—Colour-resemblances in the Cœlentera
—The Effect of Light upon the Development of Pigment
—The Pigments of the Cœlentera—Optical Colours.

THE pigments of the Protozoa are mostly very plant-like, chlorophyll and lipochromes being exceedingly common. There is often an interesting alternation between chlorophyll and red or yellow lipochrome pigments, very similar to that seen in the vegetative organs of the higher plants. Thus *Hæmatococcus* is sometimes red and sometimes green : it is always red in the resting stage, and gradually acquires the power of movement and a green colour simultaneously. Sometimes in this form and in *Euglena* there is merely a red ring round the nucleus ; when the green colour develops it begins at the periphery and gradually spreads inwards. Rostafinski found that the red colour is produced by a combination of two lipochromes, distinguished by their solubilities,

—an interesting fact, because this association or lipo-chromes in pairs is widely spread throughout both the animal and vegetable kingdoms. In *Euglena sanguinea* the colouring-matter may be absent with-out apparently any resulting specific difference. The red colour occurs chiefly in spring, in autumn, in the dry condition and in bright sunshine, a state of affairs quite comparable to that which obtains in the higher plants. According to Engelmann (1882), the red colouring-matter is capable of evolving oxygen as well as the green.

Among other Protozoa which do not contain chlorophyll, red pigments sometimes occur, often being found only in the so-called eye-spots. In mass the red colouring-matter may sometimes render the tiny organisms very conspicuous; thus Agassiz speaks of *Globigerina* occurring in floating masses of scarlet colour, and forming an appreciable factor in the coloration of the ocean-surface.

A brown colouring-matter, perhaps identical with the pigment of Diatoms (diatomin), seems to have a wide distribution among the Protozoa, but the question whether it is an intrinsic or a derived pig-ment is as yet undetermined. It occurs occasionally in the cortical layer of *Vorticella* for example.

Among the Protozoa containing chlorophyll or lipochromes we must also mention the Radiolaria. Many of these contain the so-called yellow cells, which are little masses of protoplasm apparently coloured by chlorophyll, plus some other pigment. There is strong evidence in support of the conclusion that these are unicellular algæ living in symbiosis with the Radiolarians (Geddes, Brandt). Others contain phæodia or phæodellæ, which are similar masses impregnated with fine granules of brown pigment of unknown characters.   Several suggestions as to function have been made, but none seems well established.   Karawaiew considers that they play an important part in the assimilation of food, but in what respect is not quite apparent.

Among the Foraminifera, pigments perhaps of lipochrome nature are very common.   We have already spoken of *Globigerina* as being bright red in mass; a similar pigment is described in some detail by Fritz Schaudinn in *Myxotheca arenilega*. This is a very large form, and in it the whole of the protoplasm is coloured a bright Pompeian red by means of a finely granular pigment.   The pigment is soluble in alcohol, and was found to be absent in only two cases out of a large number examined. The organism was observed to feed on Copepoda, which are often very brightly coloured organisms, and we must allow the possibility that the pigment was derived from the food.   It may be thought that this suggestion is too freely made for the colours of the Protozoa, but it should be remembered that in organisms of such great simplicity it is difficult to clearly distinguish between pigment directly introduced with the food and intrinsic pigment; in their case derived pigment has not quite the same meaning as in the case of cœlomate animals.   Even in the Cœlomata, indeed, colouring-matter introduced into the gut with food may have a direct importance in coloration; many of the transparent herbivorous worms such as Nemerteans or Annelids are coloured green by the contents of the gut.   It is likely that many of the colouring-matters of the Protozoa are as adventitious and unimportant as these, but of the physiology of the Protozoa relatively little is known.

Apart from the lipochrome pigments and chlorophyll we find that, as already seen, blue and violet pigments are not uncommon in the Protozoa. In *Stentor cœruleus* notably, the presence of a blue pigment in the alveolar layer has long been known. According to Bütschli the pigment may be blue, red, rusty-yellow, or coffee-brown. It was examined spectroscopically in 1873 by Prof. E. Ray Lankester, who found that it gave a three-banded spectrum. More recently it has been again investigated by Mr. Herbert Johnson, who has made a special study of the American species of *Stentor*.

In several of these chlorophyll is present in the form of the so-called zoo-chlorellæ, but these are entirely absent from *S. cœruleus*, which is the largest form and the one which is hardiest in aquaria. In it there is abundant pigment present in the ectoplasm, the pigment being arranged in stripes which correspond to ribs or stripes in the ectoplasm. The stripes are very inconspicuous in species in which pigment is absent, but can be demonstrated in these by differential staining. The occurrence of pigment in these simple organisms showing a differential distribution associated with morphological differences in structure and physiological differences in reaction to chemical agents, is a fact of much theoretic interest. It will be noted that the result of the differential distribution is to produce a simple form of marking —a true coloration in the common use of the term.

As to the pigment itself, according to Johnson it normally varies in tone from bright sky-blue to pale sea-green or even dull bluish-gray, but if the organisms are kept under unfavourable conditions it becomes reduced in quantity and changes to a yellowish-brown colour. This change always occurred when the Stentors were artificially divided, and Johnson never found the blue colour to be regained when once lost—a curious fact (the statement, of course, refers to forms kept in confinement). Individuals are sometimes found which are almost devoid of pigment. Schuberg observed a fact confirmed by Johnson that the pigment is thrown out of the living Stentors apparently in the same way as that in which fæcal matter is got rid of; the pigment tends especially to accumulate near the point of attachment in forms which have remained long in one place.

The blue pigment is extremely stable, not being dissolved by alcohol, ether, etc., nor attacked by acids or alkalies.

Another Protozoan (*Nassula*) also contains a blue pigment which is probably derived from the *Oscillatoria* of the food.

Another beautiful violet pigment, apparently of unknown characters, is described by Dr. O. Nüsslin in a Protozoan (*Zoonomyxa violacea*) found in the Herrenwieser Lake. This organism has its protoplasm filled with numerous small violet vacuoles, sufficiently abundant to colour the whole organism violet. The pigment has a superficial resemblance to one described by Greeff in *Amphizonella violacea*, but in that form the pigment is granular, while here it is in solution. The colouring-matter is very susceptible to the action of reagents, being destroyed by very dilute alkalies and acids, iodine or alcohol. It also disappears after death or encystment. This extreme instability is interesting, for, as we have seen, it is characteristic of so many blue or violet pigments.

In general the pigments of Protozoa seem to be usually chlorophyll, lipochromes, or blue or violet pigments of unknown relations. The notes given above are obviously incomplete, but the pigments have been little investigated. The notes may, however, be sufficient to emphasise the points of importance about the pigments of the group. These are briefly as follows : in spite of the fact that the colours show great variability, we find every now and again in the group complex and unstable pigments of vivid and beautiful tint ; these may in some cases, as in that of chlorophyll, have an important function, but such functions are entirely unknown—it would seem that in some instances the pigments are merely introduced with the food ; there is also considerable evidence to show that the pigments vary in tint in harmony with the varying physiological conditions of the organism. In all these respects the pigments of the Protozoa afford an interesting commentary on those of higher animals.

THE COLOURS OF SPONGES

About the colours of sponges it is not at present possible to say much. It is a familar fact of observation that they are exceedingly variable and often bright ; red, orange, yellow, green, and dull colours

are all common.   The green colour of the fresh-
water *Spongilla* is usually asserted to be due to
chlorophyll, but green tints occur also in marine
forms.   Sponges whose colours vary from greenish-
yellow to red almost all contain lipochromes.   The
pure red lipochrome known as tetronerythrin is
widely spread, and is often associated in sponges with
a peculiarly penetrating odour like that of ozone.
Krukenberg was of opinion that in these sponges the
pigment played some part analogous to that of
chlorophyll in plants, being of importance .in a process of assimilation.   It does not appear that his
observations have been repeated or confirmed.

In *Hircinia variabilis* and some other sponges
there is a red pigment which, according to Krukenberg, is very similar to the pigment of the Red
Algæ, and which is readily decolorised by reducing
agents.   Some of the Aplysinidæ contain pigments
of the type described by Krukenberg as uranidines,
which are of yellow colour, but tend to rapidly undergo
oxidative changes which turn them black.   In other
forms, *e.g. Chondrosia*, dark pigments occur—a fact
of some interest, because these are rare in simple
organisms.

THE COLOURS OF SEA-ANEMONES, CORALS, JELLY-FISH AND THEIR ALLIES (CŒLENTERA)

The group of animals whose colours we have next
to consider includes some of the most beautiful of
existing organisms.   Beautiful as are the changing tints
of birds and butterflies, they lack for many people
the subtle fascination possessed by the delicately

translucent colours of the Cœlentera. That some part at least of the charm is due to childish reminiscences of tales of coral islands and tropical seas, few would deny, but apart from this, many of the polypes and sea-anemones of our own shores are adorned with tints which afford intensest pleasure to a colour-loving eye.

The Cœlentera include a large number of forms, which are almost all marine, and are found in greatest abundance in warm seas. Many of them are of sedentary habit, and frequently of peculiarly plant-like appearance ; many in their method of growth or in their peculiar shape present a strong superficial resemblance to seaweed. All are characterised by relatively great simplicity of structure, and therefore, in accordance with the principles which we have already considered, their colours are due to the presence of pigment in the coloured tissues, and not to effects of structure. As a group, therefore, the Cœlentera are plant-like not only in general habit, but also in the development of a large series of brilliant pigments. Owing to their simplicity of structure and the absence of true internal skeleton, all are during life more or less transparent or translucent, and this translucency adds an extraordinary delicacy to their colouring. Unfortunately this colouring cannot be seen in preserved specimens, both on account of the fugitive nature of the pigments, and on account of the loss of the transparency during the process of preservation. During life the transparent appearance is largely increased by the habit which most possess of distending the tentacles and the body by means of sea-water, while death is usually

accompanied by a process of contraction which not
infrequently destroys all beauty of form as well as of
colour. All these causes combine to render it practically impossible, at least with present methods of
technique, to adequately preserve the large majority
of the Cœlentera. Frequenters of museums will
recall the glass models or crudely tinted chromo-
lithographs which usually fill the cases devoted to
the group. Although, therefore, the beauties of the
coral-reefs must to most of us be merely a matter of
hearsay, yet it should be emphasised that Cœlentera
are inhabitants of temperate as well as of tropical seas,
and that many a pool on the British coast will display organisms whose colouring differs in degree
only, and not in kind, from that of the denizens of
the most marvellous coral-reef.

It is not necessary here to discuss the structure
of the Cœlentera ; we need only recall the fact that
most of the organisms included in this class are either
polypes of the nature of a sea-anemone, or jelly-fish
of the familiar type, and that both these forms may
occur in the course of a single life-history. For our
purpose it is sufficient to think of a polype as consisting of a hollow column fixed at one end to some
solid body, and bearing a mouth surrounded by
waving contractile tentacles at the free end. Such
polypes may grow singly as do the sea-anemones, or
they may grow together in " colonies." The polypes
in such colonies are connected together by a fleshy
substance traversed by numerous tubes. If lime be
laid down in considerable amount in this fleshy
substance, the colony forms a " coral." This " coral "
or limy skeleton is really outside the individual

polypes, but owing to the way in which these are spread over it, they have a superficial appearance of enclosing the coral within their own soft parts.

DISTRIBUTION OF COLOURS

The coral-reefs of warm seas are largely formed of such colonial sea-anemones, the simpler colonial forms which do not produce a limy skeleton being notably absent from these regions. In their living condition on the reefs the polypes themselves, as well as the skeleton with its organic covering, are all coloured, often with the brightest of tints. Further, it seems to be relatively rare for the whole colony to be of one tint. Sometimes the polypes are sharply contrasted in colour with the coral, sometimes the youngest portions of the colony differ entirely in colour from the older ; while in those forms in which the individual polypes attain a considerable size, the tentacles may be banded or tipped with a colour quite different from the ordinary ground - colour. According to Dr. Hickson (*A Naturalist in N. Celebes*, London, 1889) the commonest tint is a deep greenish-brown, and next to that, and especially in the younger parts, a bright green. Again, a study of the descriptions and plates in Mr. Saville Kent's great book on the Barrier Reef of Australia shows that after these come shades of red, pink, and yellow, and more rarely electric blue. According to Saville Kent there is great and striking variation in tint both within the limits of a species and even during the course of the life-history of a single colony.

Although descriptions of coloured organisms do not as a rule convey much idea of their beauty, it may illustrate the problem before us if we give details of some of the corals studied by Mr. Saville Kent during his sojourn at the Barrier Reef. It is perhaps not unnecessary to repeat that one must read into the list of tints something of that purity of colour, that varying light and shade, which have made the glories of spring and autumn a never - failing source of artistic inspiration.

THE COLOURS OF CORALS AND SEA-ANEMONES

Among the masses of coral which go to form the reef, the different species of *Madrepora*, or stag's horn coral, are usually very conspicuous. In this country specimens of *Madrepora* are quite familiar as slender branching stems studded with tiny openings, but they reach us always in the white or bleached condition ; in their natural condition almost all are brightly coloured. Thus in one species (*M. prostrata*) the whole colony is usually bronze-green with yellow tips, but it may be bright green with yellow tips, or more rarely shrimp-pink with yellow tips. Again, another species is remarkable in having the ends of its branches crowned by larger cells than those which constitute the branches themselves. In this case the branches are pale yellow to white, the polypes being light brown to greenish-yellow ; the large terminal cells are a delicate china or bright turquoise blue, their contained polypes being of an emerald - green tint. Another species, *M. hebes*, shows a large amount of

variation.   Most frequently it is a dark brown colour
with white extremities ; sometimes the whole surface,
including the polypes, is a vivid grass-green.   Mr.
Saville Kent describes a large colony which when
first examined was pinkish-brown with greenish-
white growing apices, the polypes throughout being
a light emerald - green.   On revisiting the colony
after two years, it was found that the surface of the
colony was of a clear seal-brown with white or pale
lilac-blue tips, while the polypes were a clear red-
brown, the tips of the tentacles only being a greenish-
white.   This variation is exceedingly marked, but
the author seems to intimate that it was accompanied
by a retardation of the growth of the colony.   It is
unnecessary here to give further details as to the
colouring of the Madrepores ; the above descriptions,
which are taken almost verbatim from *The Great
Barrier Reef*, are sufficient to prove the statement
that these corals are remarkable for their contrast of
colour, the contrast being produced by the respective
colours of polypes and ground-substance, or by the
different tints of mature and growing regions.   The
colours mentioned are exceedingly common throughout the genus.

The same brilliancy of tint is observable among
the other corals ; the banding of the surface of
individuals is especially remarkable among the large
sea-anemones or the solitary corals.   The occurrence
of bands of colour, especially round the tentacles, is
admirably shown in many of our native sea-anemones,
as are also the essential peculiarities of colouring.
Note, for example, the green and magenta colouring
of *Anthea cereus*, the bright turquoise beads at the

base of the tentacles of the common *Actinia*, and so on. These beads are little clusters, batteries, of sting-ing-cells, and are sometimes called nematospheres. In some of the large tropical anemones the tentacles are greatly branched after the fashion called pinnate, and some of the pinnæ are modified to form nemato-spheres, which are thus borne like little fruits on the surface of the tentacles. These are especially well developed in *Heterodactyla*, and there display the most wonderful beauty of colour. Often they are a brilliant translucent violet, with an apical spot of vivid emerald-green ; in another species they are a bright amethyst with a terminal spot of a darker tint. These nematospheres occur in clusters, and when viewed under the low power of the microscope, may, according to Mr. Saville Kent, " be appropriately compared to currant-like fruit, carved out of amethyst, with a crystal of amethyst inserted, to represent the cicatrix of the antecedent flower." The occurrence of such detailed beauties of form and colour in these simple organisms is of great theoretic interest.

Among other interesting organisms included in the Cœlentera we must notice the blue coral (*Heliopora*). This obtains its name from the fact that its colour is more permanent than that of most corals, and the coloration is therefore quite marked in dried speci-mens. Curiously enough, the blue colour is more intense in the middle of the coral than on its outer surface, which is simply of a blue-gray colour. The polypes are small and pale-coloured. This form is of great interest because the special group to which it belongs, the Alcyonaria, do not now form as a rule continuous living skeletons, although many fossil corals belonged to it. The only other existing reef-building Alcyonarian coral is *Tubipora*, the organ-pipe coral; here the coral itself is of a bright red colour, and the polypes either brownish-red with green-tipped tentacles, or entirely of a delicate green colour.

COLOUR - RESEMBLANCES IN THE CŒLENTERA

We have already noted the plant-like habit of growth which causes so many of the simpler Cœlentera to be popularly mistaken for seaweed, but there are also a few detailed resemblances which have been described. Thus the species of the sea-anemone genus *Actinodendron* have much-branched tentacles which resemble the fronds of seaweeds or ferns very closely. In the case of *A. alcyonoideum* the habitat is in sandy pools in the corners of coral rocks; the long tentacles float freely in the water and sting with extreme severity when touched; the colour is dull and seaweed-like. It is suggested by Saville Kent that the resemblance to seaweed entices organisms within the reach of the deadly tentacles, and that this is therefore an example of "alluring coloration." A species of a related genus (*Megalactes griffithsi*) has tentacles of similar nature, and inhabits similar situations, but is distinguished by a relatively elaborate system of coloration, being marked by radiating white lines on a ground of lilac, pale sea-green, gray and buff. Again, in the species of *Heterodactyla*, already mentioned for their elaborately beautiful fruit-like nematospheres, the branched tentacles are grass-green in colour and "present the aspect of aggregated tufts of fine, brightly-coloured moss, or, yet more appropriately, certain varieties of the very finely divided leaves of cultivated parsley." So also one of the species of mushroom coral, *Fungia crassiten-taculata*, when in the fully extended condition, " bears a considerable resemblance to a crowded growth of the common green seaweed, *Enteromorpha*." There is, however, great colour-variation in the species, and it is only the green variety which presents this appearance. Finally, we may mention the case of the " lettuce-corals " (*Tridacophyllia*), which, especially in the green species, are said to closely resemble in their peculiar method of growth leaves of lettuce or endive. This is an interesting case of resemblance, for it would require an exceedingly enterprising biologist to construe it as a case of protective or alluring coloration.

THE EFFECT OF LIGHT UPON THE DEVELOP-MENT OF PIGMENT

On this subject there are a few interesting observations. Mr. Saville Kent observed that in *Euphyllia glabrescens* the tentacles were frequently green or brown with paler tips, but where they were completely shaded from the light they were quite transparent and colourless with faintly tinted tips. Similarly, forms like *Symphyllia* which only unfold their tentacles at night, have these transparent or colourless, while the exposed parts are coloured with the usual brown or green pigments. As a curious exception to this rule he found that a species of *Dendrophyllia*, *D. coccinea*, was always coloured a bright red, even when shaded.

Similarly, *D. ramea* also displays the same deep tint whether it grows in shallow water (7-8 fathoms) or at great depths (600 fathoms).

In connection with the effect of light we may mention the prevalence of blue colours among the pelagic jelly-fish, which many would regard as directly due to the action of light.

THE PIGMENTS OF THE CŒLENTERA

The number, beauty, and great variability of the tints of the Cœlentera make the question of the nature of their pigments one of great interest, but, as is so often the case, the bright pigments are very unstable and their examination is a matter of great difficulty. In consequence, the observations which have been made are, in most cases, very incomplete. We shall not attempt to give a detailed account of the pigments already described, but shall merely describe the characters of the better known of them.

We have already emphasised the predominance of a green tint in the sessile Cœlentera, and the frequent tendency for this tint to be in whole or in part replaced by another colour, such as blue, pink, or brown. The brightness of the tint has suggested to many the possibility that the pigment might be chlorophyll or some related colouring-matter, imparting to the organism the power of taking carbon from the air and evolving free oxygen. Among the more recent supporters of this view we have Prof. Hickson, who during his stay in N. Celebes remarked on the absence of Algæ in the neighbourhood of coral-reefs, and suggested that their green pigment might enable the coral polypes to physiologically replace the missing plants. The suggestion has the more force in view of the fact that many anemones have in their inner layer the so-called "yellow cells," which many regard as symbiotic Algæ. Further, Prof. Geddes some years ago succeeded in showing that some green anemones, *e.g. Anthea cereus*, possess in sunlight the power of evolving free oxygen. Krukenberg was unable to confirm Geddes's results as to the evolution of oxygen, but subjected the pigments of *Anthea* to a careful examination. He chose specimens of *Anthea* which were bright green with purple tips to the tentacles; and though he found that an alcoholic extract contained a mixture of pigments, he could not succeed in persuading himself that it contained chlorophyll. He speaks with considerable reserve on the question of the affinities of the pigments, and carefully guards himself against an absolute denial of the existence of symbiotic Algæ; his point simply being that there is considerable evidence against the hypothesis that *Anthea* contains chlorophyll. Observations on other pigments which have been made since Krukenberg's work, seem to justify us in speaking a little more decidedly on the subject of this pigment.

Krukenberg's observations may be summarised as follows: he obtained a pigment which dissolved in alcohol to form a solution which varied in colour from brown to green, showed distinct red fluorescence, and gave a banded spectrum; the addition of acid turned the green solution blue, and added a new band at the junction of the yellow and the green, as well as altering the position of the other bands.

These observations taken together seem to show that the pigment belongs to an interesting group of colouring-matters, which, as already mentioned, includes the pigments of the worms *Bonellia* and *Chætopterus*, of the "liver" of Mollusca and of some other animals, all of which present some superficial resemblances to chlorophyll. In view of the fact that the cells of the "liver" in Mollusca contain yellow-brown granules which certainly contain the characteristic pigment (often called enterochlorophyll), I am of opinion that it will probably be found that the so-called yellow cells of *Anthea* are merely granules of the characteristic pigment, probably mingled with some other. Dr. M'Munn (1885) on spectroscopic grounds certainly denies that the *Anthea* pigment corresponds to his "enterochlorophyll," but as the animals contain a mixture of pigments, it seems not improbable that the anomalous appearances noticed by him were due to the simultaneous occurrence of several pigments in the same solution. The resemblances to the chætopterin group are certainly very striking. Now these "yellow cells" are widely distributed among sea-anemones, and if we may assume a similar wide distribution of a pigment of the chætopterin group, it is possible to account for some of the remarkable colour-phenomena of the Cœlentera ; these may be merely the result of the properties of the dominant pigment.

The better known pigments of the chætopterin group, viz. chætopterin, bonellin, and "enterochlorophyll," show two marked peculiarities. In the first place, associated apparently with their complex

spectrum, they exhibit a lack of definiteness and purity
of tone ; in dilute solutions chætopterin is a delicate
pure green, but when concentrated it becomes muddy
and shows strong red fluorescence.   It seems not unreasonable to suppose that this peculiarity will affect
the nature of the tint in the living organism.   Thus
if the same pigment occurs in very delicate translucent tissues, and in denser and more opaque ones,
it seems probable that the characteristic differential
absorption will result in striking difference of tint in
the two cases.   May not this account for some of
those marked differences in colour between old and
young parts of a colony, between polypes and " coral,"
between tentacles and body, and so on, upon which
we have already dwelt ?

In the second place, the pigments of the chætopterin group are remarkable for their extreme sus-
ceptibility to the action of reagents, especially to
acids and alkalies.   Acids produce remarkable variations in tint corresponding in part to new pigments ;
it is probable that a considerable number of colouring-
matters can be produced by the action of different
reagents.   Now Krukenberg noticed that the colour
of sea-anemones varied according as they secreted a
tryptic ferment (i.e. one which acts in an alkaline
medium) or a peptic one (acting in an acid medium);
it seems therefore probable that the colour in any
given case will depend upon the reaction of the
surrounding tissues.   This may again account for
some of the local variations in tint, while it is not improbable that the vivid green colour so often seen may
be due to a pigment derived from the pigment of the
" yellow cells."   Chætopterin and enterochlorophyll

at least can be made to yield very bright green derivatives.

Dr. M'Munn (1885) has described by means of the spectroscope a number of other pigments in sea-anemones, but too little is known of these to make it profitable to detail here their names and properties. There is, however, one interesting pigment, called by Moseley polyperythrin, which deserves further notice. Moseley found this pigment in a number of simple stony corals and in a few anemones and jelly-fish, almost all from deep water. It is of a deep madder-brown colour, and in the case of the corals sometimes coloured uniformly both the soft parts and the coral, sometimes the soft parts only. Further, in some specimens it was uniformly distributed, while in others it occurred in streaks or was totally absent. The pigment dissolves in acidulated alcohol or in dilute acid to form a pink solution with green fluorescence, and gives a spectrum which in some respects resembles that of the pigments of the chætopterin group. It seems not improbable that this pigment is the result of the modification of a pigment allied to the " chlorophyll " of *Anthea*.

Another interesting series of pigments, apparently not allied to the preceding, are those producing the blues and browns of the surface jelly-fish. There is reason to believe that the blue colour of *Cyanea*, of the common *Aurelia*, of *Rhizostoma*, and of *Vellela* are all due to a pigment called by Krukenberg Cyanein, which is recognised by the following characters. It is soluble in water, especially in water containing neutral salts (*e.g.* sea-water). It is unaltered by weak acids, but strong acid, alcohol, benzol, chloroform, phenol, heat, etc., destroy the blue colour and give a precipitate of dull red or brown colour. The blue solution in water gives three (or two?) bands in its spectrum, but Professor E. R. Lankester (1870) did not succeed in getting these bands with *Vellela;* the reddish-brown substance apparently gives no bands. Further, in *Rhizostoma Cuvieri* the blue colour is somewhat variable; it chiefly occurs in young specimens; the older, especially when carrying eggs, are of dirty red or reddish-brown tint. This must surely be due to the modification of the unstable blue pigment under the influence of some specific change in the chemical characters of the protoplasm of the organism. Then again, Professor M'Kendrick found that the brown colouring-matter of the jelly-fish *Chrysaora* was soluble in hot sea-water, yielding a dark brown solution of acid reaction, which gave a reddish-brown precipitate with strong alkali. It seems probable that this colouring-matter is also the result of a modification of cyanein.

These cases show that in all probability the number and variety of the tints in the Cœlentera are due in large part to physical or chemical changes in a few complex pigments. As to the cause of these changes, it must be noted that the pigments are here deposited in internal structures where active metabolism is going on, and not, as in many higher animals, in superficial inert tissues. The physiology of the higher Vertebrates shows that in them there is a metabolism of pigment as active as any that can occur in coral polypes; but while in the latter every change is apparent on the surface, in the former the changes are concealed by the intervening passive tissues, and the colour has a characteristic permanence. The varying translucency and delicacy of the tissues in the Cœlentera further determines the apparent colour of the pigments, so that we may almost describe these as simple forms of optical colours. It is likely that the frequent variation of colour in simple organisms is to be accounted for in a similar fashion.

Among the more special colour-phenomena of the group, we may notice the comparative rarity of the lipochromes. They are said to occur in the red coral of commerce, in the skeleton and soft parts of some of the Gorgonidæ, probably in the *Dendrophyllia* mentioned above, and in a few others.

The colouring-matter of the blue coral *Heliopora cærulea* also deserves mention. It is very different from any other blue pigment known, and is very insoluble. If, however, the coral be decalcified with hydrochloric acid, the pigment is set free, and may be dissolved in alcohol. The solution gives no bands and turns green with alkali. It seems not improbable that the colouring-matter is metallic in origin, perhaps a salt of some metal (*cf.* the green colour of the bones of *Belone*, etc.).

From the simplicity of the structure of the Cœlentera it is naturally to be expected that true optical colours will be absent. These are not, however, quite unknown, for the calcareous axes of some of the Gorgonians display brilliantly iridescent colours. Agassiz (*Voyage of the Blake*, vol. ii. pp. 144, 145) describes the species of *Iridogorgia* as having axes of a bright emerald-green or of burnished gold, while others have a lustre like mother-of-pearl. All the iridescent forms inhabit deep water, and the colours are probably due to the same cause as the colours of shells, which also tend to be especially iridescent when found in deep water.

We have already spoken of the prevalence of phosphorescence in the group.

CHAPTER V

COLOUR-PHENOMENA IN WORMS

Colours of Turbellaria and Nemertea—Pigments of Gephyrea—
Colours of the Chætopoda, Structural and Pigmental
Colours—The Pigments of the Capitellidæ—General
Characters of Coloration of Leeches and Origin of Mark-
ings—Pigments of Polyzoa and the Origin of Pigmentation.

As central and unspecialised forms, the worms are
of considerable. interest in a comparative study of
colour, and there are several interesting facts in
regard to them which make it desirable that we
should consider the more important groups successively.

The somewhat grim associations which cluster
about the Platyhelminths do not lead us to expect
bright colours among them, and yet in point of fact
the free living forms often exhibit great brilliancy.

Among the Turbellaria pigment frequently occurs
in the cells of the epidermis, in the interstices between
these cells, or in the parenchyma of the body.
Many, such as *Convoluta*, contain in the cells of this
parenchyma the so-called symbiotic Algæ of green
or brown colour. These chlorophyll cells have no
cellulose envelope, and often contain pyrenoids.

The only detailed investigations on the pigments
of Turbellaria appear to be those of Moseley, who
made some observations on two species of *Rhyncho-
demus* found in New South Wales. Of these one
was blue and the other red, the two living together
under somewhat similar conditions. The blue
pigment was insoluble in alcohol, turned red with
acids, and then dissolved in alcohol; alkali restored
the blue colour. This at once suggested that the
red pigment was due to an acid reaction in the tissues
of the red species, and was simply a modification of
the blue. Moseley could not, however, succeed in
turning the red pigment blue with alkali, and found
that it was insoluble in acidified alcohol. He was
forced therefore to the conclusion that the two
pigments are entirely different; no further investigation seems to have been made on the subject.

It is interesting to note that in spite of the
simplicity of the Turbellarians, the coloration is not
necessarily completely uniform, but may show an
arrangement into bands and spots. Thus the species
of *Geoplana* often show a dorsal stripe of green,
orange, or purple, while the rest of the body may
show spots of blue, brown, or yellow on a dull
ground.

The pigments of the Nemertea have been even
more neglected than those of the Turbellaria, though
with less excuse. We quote the following eloquent
description of their colours from Professor M'Intosh's
monograph :—" The colours of many species of the
group are of such beauty as to attract even the
casual observer, while in this respect also they widely
deviate from their supposed allies the parasitic

H

worms. The richest purples appear on velvety skins of deep brown or black, each of the soft and mobile folds giving shades that vary in intensity and lustre. Bright yellow contrasts with dark brown, white with vermilion brown and dull pink, while individual uniformity is characterised by such hues as rose-pink, white, green, yellow and olive, the gradation of colour in the various parts of a single specimen being so subtle that enthusiasm as well as skill is necessary in the artist who sets himself to the task of faithful delineation" (*Ray Soc.* vol. xxxiv. p. 2). This description is based on the British forms only, and the tropical are said even to surpass these in splendour. In addition to the beauty of colour, we find that simple forms of marking, such as longitudinal and transverse stripes, are common. Professor M'Intosh speaks also of the "silver sheen," and the "ever-changing iridescence of the active cilia," and considers that in point of beauty and variety the Nemerteans even surpass the Annelids. Light seems to have some influence on colour, but the exact extent of the influence is apparently not determined. Thus we are told that some of the most brightly coloured live in crevices and dark corners, but that under natural conditions the colour varies according to the amount of exposure to light. In captivity *Lineus marinus* turned pale, especially in the anterior region, but others, *e.g. Amphiporus lactifloreus*, developed more pigment so that the skin became opaque and of a deeper colour. Some are transparent, and the food contained in the gut (Algæ, etc.) shines through and produces a marked and peculiar coloration. There is virtually no sexual difference in colour, but in some cases, as in *Amphiporus pulcher*, the eggs are bright red, and by shining through the thin body wall, produce a striking effect on the coloration of the female. In simple animals this primitive form of colour-difference in the sexes is not uncommon.

Bürger's recent monograph with its beautiful plates confirms M'Intosh's descriptions without adding very much new information as to colour. Bürger describes the epithelium as consisting of three kinds of cells— slender thread-like cells, interstitial cells, and gland-cells. The abundant pigment may occur in any of the three. In *Lineus* and the related forms it occurs in the gland-cells, whose secretion is often grass-green. In *Nemertopsis peronea*, on the other hand, the pigment is confined to the interstitial cells, which form two long dorsal stripes of red-brown colour. Bürger describes numerous instances of protective coloration and of marked colour - resemblances between armed and unarmed forms found living together. His figures show again the frequency of bands both longitudinal and transverse, but I have not been able to find any suggestions as to their meaning or origin. The pigments also have not yet been investigated.

The remaining flat-worms, being parasitic, display no brilliancy of colour.

THE PIGMENTS OF THE GEPHYREA

In the small order of the Gephyrea known as the Echiuroidea a very interesting pigment occurs. The curious form known as *Bonellia viridis* is bright green in colour, and was for long supposed to contain chlorophyll. The pigment is said to occur in the skin and sub-epidermic cells of the female, and in the wandering cells which partially fill the reduced body cavity in the degenerate male. It is readily soluble in alcohol, the solution being green to brown according to the degree of concentration, and displaying a blood-red fluorescence. The solution gives a beautiful and complex spectrum, which changes when acid is added. The addition of acid also changes the colour from violet to blue according to the amount added. The pigment was given the name of bonellin by Sorby, and has been studied by numerous investigators. Both Sorby and Krukenberg showed conclusively that it is not chlorophyll, but the superstition dies hard, and may be still found in many text-books. Professor E. Ray Lankester (1897) has recently re-examined the pigment, and shown that it occurs in the organism in the alkaline condition—a point of some interest. There can be little doubt that bonellin is a member of the group of pigments spoken of in the last chapter, which may conveniently be called the chætopterin group. These pigments contain nitrogen but no copper, and their function is quite unknown. The curious point in connection with bonellin itself is that it is only known in *Bonellia viridis*, although a green colour is common among the Echiuroids. A new British species of *Thalassema*, described by Professor Herdman, is of an extremely vivid green colour, the tint being somewhat similar to that of *Bonellia*. Nevertheless the pigment was found to be soluble in water, to give a one-banded spectrum, to form in alcohol a pure green solution without fluorescence, and so to offer a marked contrast to bonellin. It seems most probable that this pigment is related to some of the derivatives of bonellin. The same pigment appears to occur in smaller amount in some other species of *Thalassema* and in *Hamingia.* Both these genera are further stated by Lankester to contain hæmoglobin in the perivisceral fluid and muscles.

In the Sipunculoidea, another order of Gephyreans, integumental pigments appear to be absent, but in *Sipunculus* and *Phascolosoma* the oxidised blood contains a deep red pigment called by Krukenberg hæmerythrin. This resembles hæmocyanin in being colourless when reduced, and probably in having a respiratory function. It belongs to a group of pigments called by Krukenberg the Floridines, characterised by their solubility in water and glycerine, and their insolubility in the usual organic solvents, such as alcohol, ether, chloroform, etc., as well as by the readiness with which they undergo oxidation. This pigment is similar to the one already mentioned as occurring in the sponge *Hircinia;* it is, however, difficult to believe that it can be respiratory there.

THE COLOURS OF THE CHÆTOPODA

The colours of the Chætopoda often display great beauty, and may be structural or pigmental, or due to a combination of the two. One of the simplest forms of structural colour is that displayed by the earthworm, where the cuticle exhibits a faint iridescence due to a system of fine lines on the surface. With the greater development of the cuticle in the marine worms, there is also a greater elaboration of structural colour.

As a type of coloration in Polychætes we may ake *Nereis diversicolor*, a very common worm on our shores. This species shows considerable colour-variation, the upper surface being a pure bronze, greenish-brown, green, or pinkish, while the lower surface is bright pink or flesh-coloured, the whole body showing a very distinct metallic sheen in addition. These bright colours are most distinct in the large adult specimens found in the Laminarian zone, and fade very rapidly after death, whether the specimens are preserved in alcohol or formalin. Preserved specimens are dead-white in colour, but still show a faint sheen.

Taking such a form as a starting-point, we have on the one hand worms in which the colour is predominantly structural, and on the other those in which it is predominantly pigmental. Structural colours are especially marked in cases where the characteristic bristles attain great development. In the sea-mouse (*Aphrodite*), for example, the bristles form a dense felt-like mass exhibiting beautifully iridescent tints, which are in life much obscured by the mud with which the animals are usually covered. The beautiful golden crown of bristles which *Pectinaria belgica* protrudes from its tube, is another example of structural colour occurring in connection with specialised cuticular structures.

Colours due to pigments are also often exceedingly beautiful in these marine worms. The vivid green of *Eulalia viridis*, of *Sabella*, and of many others, the pink of *Terebella* and *Euchone rosea*, and the bright red filaments of *Cirratulus*, are examples of the bright tints which are so widely spread in the group.

Before passing on to consider what is known as to the nature of the pigments, we may note the different ways in which the external coloration of the Annelids is produced.   In the smallest and simplest forms there is no pigment either in the blood or tissues, and the animals are therefore transparent and colourless.   But when they are herbivorous, as is often the case, the contents of the food-canal may give them a green colour.   The next stage may be described as the condition when the tissues retain their transparency and lack of pigment, but the coloured blood shines through the thin body wall, and gives the animals a distinct and often bright colour.   This is well seen in the bright red *Tubifex* of ditches, the green *Sabella* in its sand-tubes on the seashore, and others.   Again, the tissues may be devoid of pigment, but may be too opaque to allow the blood-pigment to shine through except in certain regions.   Thus, for example, in *Gordiodrilus tenuis* the general colour is a cream white produced by the cœlomic corpuscles, but the body is marked by longitudinal red stripes produced by the shining through of the larger blood-vessels. Generally, however, we find that forms in which the tissues are too opaque to allow the blood to be directly . visible not infrequently develop pigment in the tissues.   Thus the common earthworm contains a certain amount of pigment scattered among the muscles; and some earthworms owe their very dark colour to pigment occurring in this situation, or present in considerable quantity in the cells of the peritoneal epithelium. Accompanying the differentiation which gives size and opacity to the body, there is frequently in Annelids, as already noticed, a differentiation of the cuticle which gives rise to structural colour. It is curious to note that, contrary to the usual rule that bright colours in worms when not due to structure are the result of the shining through of coloured internal structures, we find that the bright colours of the little Oligo-chætes belonging to the genus *Æolosoma* are said to be due to coloured oil-globules contained in the skin (see Beddard, *Ann. and Mag. Nat. Hist.* vol. ix. pp. 12-19). The oil-globules are in the different species blue-green, yellow-greeen, or orange-red, and may be of lipochrome nature.

As to the pigments themselves, if we bear in mind the statement just made that, when bright, the colours of worms are due in the general case to coloured in-ternal tissues, it must be obvious that the pigments are usually those which by hypothesis are of direct physiological importance. Hæmoglobin, which is widely and irregularly distributed in the group, is one of these. It is often exceedingly conspicuous in the more delicate forms; but even in the larger and more opaque worms it may shine through the thin-walled gills or tentacles. The green pigment which occurs in the blood or cœlomic fluid of *Sabella, Branchiomma, Spirographis,* and *Siphono-stoma,* is another example. This pigment has been accredited with important respiratory functions. It was called chlorocruorin by Professor E. Ray Lankester, who stated that it is capable of existing in an oxidised (green) and a reduced (red) condition. Krukenberg denies this and regards the appearances observed by Professor Lankester as due to the marked dichroism of the green solution. The solution exhibits a two-banded spectrum, and the pigment is destroyed by strong acid or alkali.

In connection with the supposed respiratory function it is perhaps worth notice that, according to Paul Langerhans, the colour of the blood in *Sabella variabilis* may vary from a clear yellow-green to a dark brown ; while it is an old observation that in worms in general the colour of the blood varies much, being very frequently different in the various species of a genus.

Another interesting green pigment is that which, as we have already noticed, colours the aberrant worm *Chætopterus*. This pigment is of greenish colour, and is confined to the walls of the anterior part of the alimentary canal. Like bonellin and " enterochlorophyll," it gives a complex, exceedingly beautiful spectrum which changes on the addition of alkali or acid, and it in other respects shows a relation to these two pigments.

Another green pigment of brighter tint but apparently of simpler nature occurs in *Eulalia viridis*, and probably in other forms. In *Eulalia* itself it is apparently intermixed with a yellow pigment, but in the eggs occurs pure. It is soluble in water and alcohol, turns slightly blue with acids, and is destroyed by strong alkali. It closely resembles the green pigment of *Thalassema* already mentioned.

The other pigments of the bristle-bearing worms are not well known ; it is possible that some of the pink pigments are lipochromes.

As to the colours of Polychætes in general, the striking features are the presence of optical colours, the number of green forms, and the absence of those beautiful and elaborate markings which are so characteristic of the leeches. The plan of the coloration is throughout much simpler ; when brilliant, it seems to be usually dependent upon coloured internal tissues shining through the skin. A great number of forms, moreover, especially those inhabiting tubes, are colourless and transparent.

The colours of the Oligochætes, such as the earthworm, hardly merit separate notice : they are mostly dull or inconspicuous, but in some cases show a tendency to develop simple patterns. Of this tendency our own brandling (*Lumbricus fœtidus*) affords a convenient example.

THE PIGMENTS OF THE CAPITELLIDÆ

The pigments of the small group of Polychætes known as the Capitellidæ are perhaps worthy of more detailed notice. The Capitellidæ are the subject of one of the large Naples monographs, and the author, Professor Eisig, makes some observations on their colours which have been very widely quoted. The Capitellidæ are remarkable among Annelids in having no closed blood-vascular system, the blood being contained in the general body cavity. According to Eisig, it always contains hæmoglobin, and is of a bright red colour which directly affects the superficial coloration.   The deposition of pigment either in the cuticle or in the hypodermis is rare, but in *Capitella* and *Heteromastus* granules and droplets of yellow-brown pigment lie between the cuticle and the hypodermis in patches.   As the nephridia contain similar granules, Eisig is of opinion that these pigment patches are due to substances excreted by the nephridia which do not reach the surface but, lying in the skin, are got rid of at the (hypothetical) moult.   In *Capitella*, in the head and tail regions, the skin is of a red-yellow colour, which is due to clusters of blood-discs containing excretory particles again lying between the cuticle and hypodermis.   According to Eisig these blood-discs, after they have taken up excretory particles, lose their power of circulating and stagnate at the areas in the skin mentioned above.   Further, Eisig considers that the pigment is directly derived from the hæmoglobin of the blood, that it occurs in the nephridia in association with guanin as one of the nitrogenous waste products of the organism, and that it may find its way to the skin and bristles and there be an important agent in coloration.   He therefore concludes that pigment is always, or at least frequently, in origin a waste product eliminated by the skin. This is of course a position which has been advocated or supported by many writers, and that such a utilisation of waste product does occur has now been abundantly proved, *e.g.* in Lepidoptera ; that it occurs in Annelids is eminently probable, and is rendered more so by Graf's observations on leeches, to which we shall afterwards refer.   At the same time it is perhaps permissible to remark that it does not appear that any good end is served by the over-hasty application of the term waste product to all pigments occurring in the skin, even when such pigments are periodically eliminated by means of a moult.   In human physiology the term waste product has a perfectly definite meaning, and it is surely desirable in the interests of scientific nomenclature that a term implying a certain chemical composition should not be loosely applied to unknown substances. Then again the occurrence of a pigment in a structure which is periodically cast and renewed is not absolute proof that the substance is useless or noxious. Thus the cuticle of the Crustacea certainly contains proteid, and, according to Krakow, also glycogen, and glycogen is certainly used up in its formation, but proteid and glycogen are not waste products.

THE COLORATION OF LEECHES AND THE ORIGIN OF MARKINGS

With regard both to their pigments and to their coloration, the leeches, in the case of the more specialised forms at any rate, are sharply contrasted with marine worms.   The cuticle compared with that of many worms is unspecialised, bristles are absent, as are also structural colours.   The smaller forms may exhibit little pigment, but the differentiated forms, like *Hirudo*, are characterised by the development of a large amount.   This pigment is not uniformly distributed, but is arranged, especially on the upper surface, in definite lines and spots, which give rise to a beautiful and complex style of coloration.   Any one who doubts the propriety of the word

" beautiful," as applied to it, is advised to examine the
plates illustrating Whitman's *Memoir on the Leeches of
Japan*, where the native draughtsman has delineated
the colouring with a fidelity which Western artists
can only envy and admire.

The pigments of leeches are largely of the dark,
insoluble, and little-known type, but it is probable
that lipochromes are also often present.   What part,
if any, the hæmoglobin of the blood takes in coloration is unknown.

On the origin of the pigment and markings of
leeches there is an exceedingly interesting paper by
Dr. Arnold Graf.   According to this investigator
there are in leeches certain migratory cells, comparable to the yellow cells of the earthworm, which
arise from the endothelium of the body cavity,
receive waste products from the blood-vessels, and
carry these to the nephridia and so to the exterior.
The waste products received by these " excretophores " are of the nature of fine dark granules,
which are capable of acting as a pigment.   Graf
finds that all the excretophores do not reach the
neighbourhood of the nephridia, but that certain of
them penetrate the musculature of the body, and
come to lie immediately beneath the epidermis,
where the contained dark-coloured granules give rise
to surface coloration.   He further states that the
number of the excretophores and the intensity of the
pigmentation increase with age, and when, as in
albino varieties, pigment is scarce, the excretophores
are also greatly reduced in number.   The amount of
pigment appears to depend upon the intensity of
metabolism, being greatest in the most voracious

forms. When fed with food coloured with carmine after a previous fast, it was found that the carmine could be traced in the excretophores, in the nephridia, and in the pigment cells beneath the skin.

In addition Dr. Graf makes some very ingenious suggestions as to the origin of the characteristic markings of the different species of leeches. According to him these depend primarily upon the arrangement of the muscles of the body-wall; or, more exactly, the visible coloration depends upon the amount of resistance which the tissues offer to the passage of the pigment-containing cells, the muscles being the most important of the tissues concerned. The muscles of the leeches are arranged in three layers, which from without inwards are the circular, the diagonal, and the longitudinal layers. Each of these layers consists of bundles of muscle-fibres, the number of the muscles varying in different leeches. According to Graf the pigment-containing cells can pass outwards only in the spaces between the bundles, so that the coloration depends upon the number of these bundles, and this varies in the different forms. Thus *Nephelis quadrostriata* has five well-developed bundles of longitudinal fibres on the dorsal surface, the circular and diagonal muscles being less developed, and we find that it has four well-marked longitudinal stripes, corresponding to the spaces between these muscles. In *Clepsine hollensis*, Whitman, on the other hand, the longitudinal bundles are very numerous and relatively weak, while the circular are well developed; and here the surface is spotted rather than striped, this being supposed to be due to the stopping of the pigment cells at regular intervals by the strong circular muscles, so that the stripes which would be formed at the spaces between the longitudinal muscles are interrupted.

These suggestions are exceedingly interesting and ingenious, and of obvious importance in relation to the origin of markings. In view of the origin of the pigment from the blood, it would be very interesting to know whether it is in any way derived from hæmoglobin, and also whether that transfer of pigment from the gut to the surface which occurs when the leech is fed with carmine, ever occurs during the ordinary course of affairs.

The process described by Graf presents some interesting analogies to that described by Eisig for the Capitellidæ. An inquiry into the causes which determine the marked differences in colour between the leeches and the Chætopoda would be very interesting, but does not appear to have been attempted.

THE PIGMENTS OF THE POLYZOA, AND THE ORIGIN OF PIGMENTATION

In connection with the colours of worms there are a few points about the pigments of the Polyzoa which are of interest.

The Polyzoa are aberrant worms, which form colonies consisting of numerous polypides embedded in cells or chambers, which, being united together, constitute the cœnœcium or substance of the colony. The cœnœcia of *Flustra*, the sea-mat, are very familiar objects on our own shores, where they are often mistaken for seaweed. Each cell or chamber with its contained polypide is known as a zoœcium.

As to structure, we recognise in each zoœcium the firm coating which surrounds and protects the contractile polypide. The polypides themselves have a ring of tentacles surrounding the mouth, and a well-developed alimentary canal. The cavity of the zoœcium is largely filled up with the so-called funicular tissue, which consists of a network of branching cells, the meshes of the network containing numerous transparent connective-tissue cells, which may be called leucocytes.

In certain forms, *e.g.* in *Bugula neritina*, this funicular tissue is coloured by pigment which varies in tint and gives the colony a purple or yellowish-brown colour. In other cases this tissue is quite colourless. Krukenberg studied the pigments in the case of *Bugula neritina*, and found that there was in the first place an interesting colouring matter readily soluble in cold water and glycerine, to which it gave a rose-red tint. On account of the presence of this pigment, dying specimens of this species, as was observed by Cohn, coloured the water in which they were found a deep purple. When the purple solutions were kept for some time the tint changed to a pale yellow, but the original purple could be restored by shaking with air. The purple solution gave two bands in its spectrum, which disappeared completely when it turned yellow. On account of the readiness with which it could be oxidised and deoxidised, Krukenberg credited the pigment with respiratory importance. Besides the red colouring-matter there is also present in *Bugula neritina* a yellow pigment apparently of lipochrome nature.

Besides these observations of Krukenberg's on

the characters of the pigments, there are some interesting facts connected with the pigmentation disclosed by a series of experiments made by Mr. Sidney F. Harmer.   Mr. Harmer's experiments were made with a view to determine the exact nature of the so-called "brown bodies" of marine Polyzoa, which have been supposed to be excretory.

These brown bodies arise roughly in the following way.   Each polypide in the colony is only capable of a relatively brief existence ; after a certain period the organs, and especially the alimentary canal, become degenerate and, fusing together into a mass, form the so-called brown body which is found lying inside the zoœcium.   The brown colour is due in large part to masses of dark pigment which occur in the cells of the alimentary canal.   Contemporaneously with the formation of the brown body there occurs in many forms a process of regeneration, which results in the formation of a polypide bud, and ultimately of a new individual.   The brown body is either eliminated by means of the gut of the new polypide, or is simply stored up in the cavity of the zoœcium.

In studying these processes Harmer made a series of experiments with solutions of various pigments, such as indigo-carmine and Bismarck-brown.   His method was briefly as follows :—living colonies were placed in sea-water containing the pigments in solution ; after a short time they were transferred to clean water and the distribution of the pigment in the different tissues carefully studied.   The results obtained differed considerably according to the pigment employed, but in general terms it was found

I

that the leucocytes of the funicular tissue, the so-called hepatic cells of the alimentary canal, and to a less degree the cells of the funicular tissue itself, all took up the pigment. When the polypides began to degenerate, which perhaps occurred sooner on account of the treatment, the leucocytes became more or less aggregated round the brown body, while the pigment of the alimentary canal took a direct part in the formation of the brown body. The new-formed polypide is entirely devoid of artificial pigment, which is either stored up within the zoœcium or in part eliminated with the brown body. The introduced pigments are thus eliminated from the living tissues, the active agents being the three sets of cells already named. The funicular tissue took up pigment most readily in the case of *Bugula neritina*, the species in which it is normally pigmented.

Now as the alimentary canal of very young individuals shows no pigment in its cells, and as this pigment subsequently appears in gradually increasing amount in cells which certainly excrete introduced pigment, Harmer is of opinion that the view that the formation of the brown body is an excretory process is well founded, and that the brown pigmentary substance is a waste product, or at least a useless substance. The view is further confirmed by the fact that indigo-carmine, when introduced into the body cavity of other animals, is excreted by cells which are certainly excretory in nature.

It almost seems, however, as if we might go farther than this. It appears that the leucocytes in Polyzoa have not any power of taking up pigment from the so-called hepatic cells of the gut and carrying it outwards, as they have, for example, in the Capitellidæ, where carmine introduced into the gut is carried outwards to the skin by the action of leucocytes. Is it too daring a suggestion that the mysterious process of regeneration in the Polyzoa is due to the want of such a mechanism, and the consequent choking of the important "hepatic" cells with noxious substances? The drastic measure of total reconstruction may thus be compelled by the poisoning of essential cells.

This discussion may seem somewhat irrelevant to our main subject, but it is in reality of much importance in connection with theories as to the origin of pigment. In Polyzoa the cells of the gut are deeply pigmented with an apparently useless substance, and there is no mechanism by which leucocytes can carry away this pigment. In many animals the cells of the gut, or the hepatic cells, contain pigment, and the leucocytes have the power of removing pigment from these cells to the skin or to the exterior ; are we therefore justified in regarding such skin pigments as waste products? This is a question which we shall have to consider in some detail later. The suggestion is one which has been repeatedly made under various forms, and it is bound up with some interesting questions in relation to colour problems.

It should be further noticed that the purple pigment of *Bugula neritina* disappears for a certain period during the development of the young polypides, and then reappears later. Harmer differs from

Krukenberg in considering that it is nutritive in function, apparently regarding it as of importance in the nutrition of the young individuals. This case is of some interest, because it is so common to find coexisting with the so-called excretory pigments others whose function is unknown or doubtful, but which are apparently not excretory.

It is also of interest to notice that the so-called respiratory pigments, those which are capable of repeated oxidation and deoxidation, frequently occur in organisms where it is exceedingly doubtful whether they have really respiratory significance.

CHAPTER VI

THE COLOURS OF CRUSTACEA AND ECHINODERMA

Nature of Colours of Crustacea—Colour-variation—Pigments of Crustacea : (1) Lipochromes, (2) Soluble Blues—Relations of Two—General Characters of Pigments—The Colours of Echinoderma : (1) Star-fishes, (2) Brittle-stars, (3) Sea-urchins, (4) Crinoids, (5) Holothurians—Pigments of Five Classes—Relations to those of Crustacea.

WE shall in this chapter discuss both the Echinoderma and the Crustacea—not that there is any morphological affinity between them, but because there is considerable resemblance as regards pigments. In both, moreover, there is a tendency for lime salts to be associated with the coloured regions.

NATURE OF COLOURS OF CRUSTACEA

The coloration of the Crustacea presents many points of very great interest ; the beauty of the colours, their adaptation to the surroundings, and their great variety combine to render the subject peculiarly attractive.

It is unnecessary to dwell upon the general

characters of the Crustacea, but it may be well to
mention some points which are of importance in
reference to the coloration.   In the first place, we
may note that the Crustacea, like all other Arthropods, have a firm cuticle of chitin.   This chitinous
coat is not infrequently impregnated with lime salts,
and is in most cases pigmented.   The subjacent
epidermis almost always contains pigment, either in
special contractile chromatophores, or in solution in
the general cells, or more usually in both.   In some
cases, as in the adult lobster, the epidermis is so
completely concealed during life beneath the thick
cuticle that it is of practically no importance as
regards coloration, while in other cases the cuticle
is thin and translucent, and it is the epidermis which
is important in the production of surface coloration.
The degree of calcification is the most important
factor in the production of opacity in the cuticle, so
that in general terms we may say that the shell
or cuticle tends to be transparent in small forms, in
those inhabiting fresh water, and in abyssal forms.

COLOUR-VARIATION IN CRUSTACEA

As a class the Crustacea are remarkable for the
brilliancy of their colours, blue, green, shades of
orange, pink, red, and brown being all common
tints.   Nor are the bright colours confined to the
higher forms ; the tiny *Daphnia* forms vivid scarlet
patches on the surface of the sea, *Diaptomus bacillifer*
forms red patches in fresh-water lakes, and so on.
As a general rule, however, the colours harmonise
with the surroundings, this being especially true of

the smaller and more delicate forms. The resemblances between shrimps, sand-hoppers, shore-crabs, etc., and the localities which they respectively haunt are too patent to need emphasis, and phenomena of this nature are exceedingly common in the group. In many cases the colours vary in harmony with the environment, but this can only occur in forms in which the epidermis is the important agent in coloration, and is then due to the sensitiveness of the chromatophores. We find this in the shore-crab in the young stage, and in many of the small sessile-eyed Crustacea, such as *Idotea* and *Caprella*. The prawn *Hippolyte*, according to M. A. E. Malard, is green on green seaweed, brown on *Fucus*, red on red seaweed, and transparent among *Antennularia* and *Sertularia*. Experiment in artificial environment showed that this form is red in darkness, emerald-green in bright light, and brown in semi-obscurity, which seems to prove that it is the intensity of light rather than its colour which is of importance in effecting colour-change. Pouchet found that prawns (e.g. *Leander serrator*) turned brownish-red in vessels with black bottoms, and if then transferred to white vessels became yellow, passing through a stage in which they were bright blue. Pouchet stated that the brownish-red colour was due to a combination of a bright red colour due to chromatophores and a soluble blue pigment. When the prawns were removed to a white dish, the chromatophores slowly contracted and allowed the blue colour to become apparent. As the contraction proceeded the blue pigment gradually disappeared, and the ultimate yellow colour was produced by the contracted chromatophores only. This case is exceedingly curious, but is confirmed in a striking manner by an observation made on a prawn taken by the *Albatross* during deep-sea dredgings off the coast of Mexico. This prawn (*Benthesicymus tanneri*) is usually of a deep blood-red or crimson colour. In one specimen, however, the abdomen was marked with spots of blue on the second, third, and fourth abdominal segments. The spot on the second segment was partly blue and partly yellow, the line of demarcation between the two colours crossing the segment obliquely so as to produce a strikingly unsymmetrical type of coloration. In view of Pouchet's observations mentioned above, Mr. Faxon in describing the Crustacea of the *Albatross* suggests, reasonably enough, that this peculiar appearance is due to a change induced during the passage from the ocean depths to the surface. This suggestion leads on to the curious fact that the blue and green pigments are almost invariably absent from deep-sea Crustacea, which are usually shades of red, often deep red, or pink, but occasionally yellow or dead-white. When blue or green colours occur they are almost always confined to the eggs. As might be expected, many have sought to explain this fact as due directly to the absence of light at great depths. Without entering at this stage into details on the subject, it may be noted that such facts as that the fresh-water crayfish occasionally appears in a full blue or in a red variety, that the common lobster is sometimes red in the living condition, that in Copepoda living under similar conditions the eggs are sometimes red and sometimes blue, and so on, suggest that the phenomenon is not absolutely dependent upon a single environmental factor for its manifestation. The other suggestion that the absence of blue or green colours among deep-sea Crustacea is due to adaptation, because such colour would be invisible in the dark abysses, is even less satisfactory as a complete explanation of the common colour-variations of the Crustacea, for there are many facts besides those mentioned which suggest that there is a necessary relation between the different colours. The constant tendency visible in the group to oscillate between red and blue can hardly be explained throughout by adaptation. As facts are always more convincing than general statements, we may add to the examples mentioned above an account of the seasonal colour-change in *Holopedium gibberum*, one of the Daphnids, which shows clearly the relation existing between red and blue.

This little form was studied by Professor Anton Fritsch in the ponds of Bohemia, and was found to be colourless in winter. Towards the end of May the first trace of bright colour appeared in the shape of a diffuse rose-colour, accompanied by a blue tint in the neighbourhood of the mouth and two narrow blue stripes at the sides of the abdomen. The diffuse rose-colour was most common in dead individuals lying at the bottom of the pond, but it also occurred in living forms.

By the end of July brightly coloured individuals were very numerous; in these the under surface of the food-canal from the mouth to the end of the abdomen was coloured blue, except at the base of the third pair of legs, where there was a cluster of bright red cells. Only about 3 per cent had red spots on the shells.

At the beginning of August the red spots on the shell had completely disappeared, but the blue colour with the two red patches still persisted. By the middle of August the great majority of the individuals had lost their bright colours, and by September the minority was very much smaller. The few which still retained any colour had the abdominal stripes of a greenish-blue, with greenish spots at the bases of the legs, at points which had previously been red. The red spots on the shells in the early stage consisted of red and blue cells intermixed. According to Fritsch there is no reason to suppose that the "decorative colours" have any sexual significance.

This description shows very clearly how often blue and red colours are locked together, and how red may disappear to be replaced by blue or green. Quite similarly Dr. F. H. Herrick, in describing the development of the American lobster, notes the striking colour-variations among individuals, some "being bright red, others greenish-blue, and others pale blue or nearly colourless."

Illustrations of this relation between red and blue might be multiplied to any extent, and yet it should also be noted that there is not infrequently much constancy of coloration, at least in detail. The lower surface of the penultimate segment of the chelæ in *Astacus nobilis* is always a bright orange-red, although the tints of the other parts of the body may show considerable variation. Those who are in the habit of dissecting *Nephrops norwegicus* must have been struck by the constant bright red colour of the gullet, remarkable in an animal generally distinguished as compared with its allies by a marked deficiency of pigment. In general it may be said that the Crustacea exhibit a marked tendency to vary in colour, and especially to oscillate between shades of blue and green on the one hand and of red and yellow on the other; they at the same time often exhibit much constancy in detail, and in deep-sea forms the blue and green colours tend to disappear.

So far we have considered the two sets of colours as if they were entirely distinct from one another, but it is familiar to all that the lobster turns red when it is boiled; in other words, the action of a large number of agents, such as heat, alcohol, ether, dilute acids, etc., upon the blue or green series is to convert them into the red. The blue or cyanic series occur, as we have seen, in solution, while the reds occur in fixed anatomical elements — the chromatophores. This difference in distribution has caused many to contrast the two sets sharply, and to suppose that the reddening of the lobster is due to the total destruction of the blue pigment, which allows the red to become visible. By others, and notably by Krukenberg, it has been maintained that the blue pigment is a lipochromogen, which very readily undergoes decomposition and then becomes converted into a lipochrome. The following description is based on observations of my own in the case of *Astacus*, *Homarus*, and *Nephrops*; for details reference should be made to my paper on the subject.

THE PIGMENTS OF CRUSTACEA

1. *The Lipochromes.*—That the yellow, orange, and red pigments of the Crustacea are lipochromes was long since proved by the researches of Maly, Krukenberg, and others. Further, the lipochromes have been described as usually occurring in pairs—a red and a yellow together, which can be separated by the process of saponification.

A superficial examination of the colours of *Nephrops norwegicus*, the Norway lobster, shows that the epidermis is bright red, and the shell orange-red, the colour of a boiled lobster. If, however, the shell be decalcified with dilute acid, it entirely loses its orange colour and becomes dull red—the colour of the epidermis. The interest of this change lies in the fact that while the boiled lobster, and those parts of the lobster's shell which are not blue during life, are of a similar orange colour, the shells of deep-sea Crustacea tend to show such colours as " blood-red " " deep crimson," "crimson-red," and so on, and are usually not orange. Now the red lipochrome of the lobster or crayfish when removed from the shell very readily forms a combination with lime which is of orange colour. The shells of the deep-sea Crustacea contain little or no lime ; it therefore seems to me possible that the difference in colour between, say, *Nephrops* and one of the deep-sea Crustacea is not due to any difference in pigment, but only to the fact that in the former the lipochrome occurs associated with lime, and in the latter in the pure state. This is interesting, because Moseley during the voyage of the *Challenger* remarked on the fact that a bright red pigment (which he called crustaceorubrin) should occur in small surface forms like *Daphnia*, and then reappear in the ocean depths. The probability is, however, that this pigment is very widely distributed in the Crustacea, and that its apparent colour depends upon the conditions under which it occurs.

As to the simultaneous occurrence of red and yellow lipochromes in the Crustacea, there is no doubt that by various agents, and especially in the presence of heat, both red and yellow pigments can be extracted from the skin and shell of the lobster and crayfish. I am, however, much inclined to doubt the existence of both these pigments in the living condition. Certainly the yellow, if it exists, has practically no effect upon the coloration. The epidermis of the lobster yields to water a beautiful bright red solution, with no trace of orange, and does not on filtering leave behind any orange or yellow pigment. Lipochromes as such are of course not soluble in water, but the red ones dissolve very readily in solutions containing albumen.

The red lipochromes, when in solution in water containing proteid, are precipitated with the proteid on boiling, but if alkali be added to the solution, alkali-albumen is formed, and they are not precipitated even on boiling. The fact that the red lipochromes thus form compounds with alkalies which are soluble in albuminous solutions, is one of considerable interest to which we shall recur.

2. *The Soluble Blue Pigments.*—These unstable pigments constitute the cyanic series of Pouchet, the lipochromogens of Krukenberg. They occur

in the epidermis of *Astacus* but not of *Homarus* or *Nephrops*, and in the shells of the two first but not the last. From the epidermis of *Astacus* the blue pigment can be readily dissolved by water, or better, a dilute saline solution. From the shells of the lobster or *Astacus* the blue pigment can be extracted by treatment with a dilute solution of ammonium chloride, or better, with very dilute hydrochloric acid. The solution from the epidermis is a very bright Prussian blue ; that from the shells is usually paler, as it is more difficult to obtain a concentrated solution in their case. We cannot here discuss all the properties of this blue solution. It may be sufficient to say that the blue colour is very fugitive, and that the solution invariably contains traces of albumen. Agents such as heat, acid, alcohol, etc., which destroy the blue colour, turn the solution pink. The pink colour is due to the presence of the red lipochrome, which can be so readily extracted from the epidermis. From the reactions of the blue solution I have come to the conclusion that the blue pigment is a compound of the red lipochrome with a complex unstable organic base perhaps derived from the muscle, and possibly of the same nature as the so-called muscle extractives. The usefulness of dilute saline solutions in obtaining the blue pigment is probably due to the fact that these solutions dissolve out some proteid, and the lipochrome compound is soluble in these solutions (see above). The compound is readily destroyed by various reagents and then the red lipochrome reappears.

Next as to the *green* pigments of the Crustacea.

Green colouring - matters occur sometimes in the shells of various Crustacea, but also very commonly in the eggs both of shallow and deep water forms. The ovarian eggs of the lobster are a bright green, while the extruded eggs are a dark green.   Both yield a turbid green solution to water, which clears up at once on the addition of ammonium chloride, probably because this dissolves up proteid.   If the clear green solution be allowed to stand, it sometimes deposits orange-coloured drops of oil and then becomes a clear blue.   The blue solution gives all the characters of that obtained from the carapace. There is therefore reason to believe that the green colour of the eggs is due to a combination of the blue lipochrome compound found in the shell and a yellow pigment dissolved in fat.   The association of this yellow pigment with yolk in eggs is, of course, exceedingly common.

An interesting confirmation of the view here propounded as to the origin of the green colour of lobsters' eggs is found in an observation by Mr. Chiyomatsu Ishikawa.   In studying the development of *Atyephira compressa*, this author noticed that the ova in the very young stages were pale blue, but as the yolk developed they became green. It would thus seem that we are justified in saying that the green pigments of Crustacea are at least in some cases produced by a mixture of the blue lipochrome compound and a yellow pigment.   From the changes which the shell of the lobster undergoes as the blue colouring - matter is removed, I am further inclined to believe that the brown colours which are not uncommon in the group are similarly produced by a mixture of the blue compound and unaltered red lipochrome.

General Characters of Pigments

Looking at the pigments of Crustacea in general, it is seen that the red lipochromes form the central pigments of the group. There occurs in addition a widely spread yellow pigment, which does not apparently give the lipochrome reaction, and whose relation to the red remains doubtful. It is also uncertain whether some of those changes from red to yellow which have been described, are or are not due to an actual change of the red pigment into the yellow. On the other hand, there appears no doubt that the red pigment in itself or in its modifications is instrumental in the production of most of the colours of the Crustacea. When the shell contains little lime or is very thin, the red pigment present in the shell and in the underlying skin gives rise to bright red or scarlet tints. When the shell contains much lime it is often of an orange tint, and is then possibly coloured by a combination of the red lipochrome with lime. It is at least certain that the lipochrome does form orange-coloured combinations with lime, soda, etc., and the colour and insolubility of the pigment in the orange-coloured shell of, e.g., *Nephrops*, suggest the presence of such a combination there. This suggestion seems also to explain the brightness of the tint in deep-sea Crustacea where lime salts are virtually absent, and in recently moulted specimens of the edible crab before the development of the lime.

Again, the red lipochrome is apparently capable of uniting with a complex organic base to form a blue compound, readily soluble in solutions containing albumen, which probably gives rise to the blue colours of many Crustacea. When mixed with the yellow pigment or with the red lipochrome, this blue pigment gives rise respectively to green and brown colours.

It does not seem as yet possible to explain the absence of the blue and green colours from many Crustacea, and especially from those inhabiting deep water, but it may be noted that their continued persistence in the eggs is not very remarkable. Yolk is distinguished for the number of complex substances which it contains, especially such bases as neurin; it may be that it is something of this nature which forms the lipochrome compound.

From this description it is obvious that, in spite of the multiplicity of tints in the Crustacea, there is much uniformity of pigments—the lipochromes being fundamentally important in coloration. It is of some interest to note that the pigment called "entero-chlorophyll" occurs at most only in very small amount and infrequently in connection with the digestive gland of Crustacea. Dr. M'Munn has described it there in some cases, but seems to have never obtained the complete spectrum; in many cases it is certainly absent, and is probably never of great importance.

THE COLOURS OF ECHINODERMA

The colours of Echinoderms are almost as brilliant as those of the Crustacea, and are often of similar tints.   Some at least of the pigments are also similar, but those of the Echinoderms seem to be far more numerous.

1. *Star-fishes.*—Among these yellow, orange, or red colours are exceedingly common.   In an account of the deep-sea Asteroidea collected by the " Investigator," Dr. Alcock gives the colours in the fresh condition of 25 forms.   Of these 18 were of some shade of pink or red, 1 was jet black, 1 gray, 1 brown, 1 orange, 2 red and yellow combined, 1 yellow and brown.   The lipochrome colours thus predominate very largely.   Out of the 25, 11 came from depths exceeding 1000 fathoms, and all of these were of some shade of red or pinkish-orange.   There is thus some evidence to show that in star-fishes, as in Crustacea, the lipochrome colours are conspicuous at great depths, while shallow-water forms tend to display greater variability of tint.   The same statement can also apparently be made of brittle-stars, but the star-fishes do not usually extend to the depths occupied by the former.

Among shallow-water star-fishes we may mention *Linckia lævigata*, in which the upper surface is a bright Antwerp blue, the tube feet being chrome yellow. This species inhabits the waters of the Barrier Reef of Australia.   Another species, *L. milaris*, also of an intense blue colour, is found among coral reefs in the Malay Archipelago.   It is mentioned by Kükenthal as having upon it a parasitic Mollusc— *Capulus crystallinus*—one of the bonnet-limpets, which is of exactly the same blue colour.   One can only regret in these cases that exact chemical observations on the spot were impossible.

2. *Ophiuroids.*—The brittle-stars resemble the star-fishes in displaying very considerable brilliancy of colour. According to Agassiz's account the shallow-water forms exhibit the greatest variety of colour, being blue, green, red or yellow, while deep-sea forms are more usually bright orange or red. Agassiz (ii. p. 133) makes the interesting observation that the colours of the deep-sea forms fade much more rapidly in alcohol, than those of the denizens of shallow water. This strongly recalls the conditions already emphasised for Crustacea, and is probably due in the same way either to a combination between the pigment and lime, or to the want of penetration of the alcohol into lime-containing tissues. The brittle-stars not infrequently resemble their surroundings in colour; the common " Sand-stars " of our own shores are good examples of this, and various observers speak of the resemblances in colour between the forms living among gorgonians and corals, and their organic surroundings. This is, however, in curious contrast with a statement made by Kükenthal, to the effect that forms living in the interstices of coral-reefs are well protected by their surroundings, do not need protective tints, and are therefore "mostly of a blackish colour."

3. *Sea-urchins.*—In these we have the same difference in permanence between the colours of the deep-sea and shallow-water forms as in the brittle-stars. The colours seem to be usually reddish, some are violet or claret-coloured, others brown or orange. A form (*Diadema setosa*) which is found on coral-reefs has five bright ultramarine blue spots arranged round the aboral aperture, while *Astropyga freuden-*

bergi has similar blue spots arranged in radiating lines. In *Diadema* the spots have been called eye-spots.

4. *Crinoids.*—The Crinoids are remarkable in showing great individual variation in colour. During life the colours are usually brown, red, purple, violet, green, yellow or white, but the curious form called *Holopus* is of a jet-black colour. Agassiz remarks that the colours of the deep-sea and shallow-water forms show remarkable similarity.

As to individual variation in colour Moseley notes that during the voyage of the *Challenger* he found specimens of *Pentacrinus* which were purple, and others which were yellow or pale-coloured. Agassiz noticed the same thing in both *Pentacrinus* and *Rhizocrinus*, while Malard describes it in greater detail for *Comatula*. Malard found a large number of specimens of the rosy feather-star on a buoy near La Hougue which were of three distinct colours, violet-red, orange-red, and white and red with reddish pinnules. Clinging to the feather-stars, Malard found numerous specimens of the Crustacean *Hippolyte*, and he states that "at least in the majority of instances" the prawns resembled their neighbours so much in tint that it was exceedingly difficult to distinguish them. Malard regards this as an ordinary case of colour resemblance, but it is also interesting on account of the numerous analogies which exist between the colours of the Echinoderms and the Crustaceans. We find in the two groups not only similar colours, but also a similar range of variation. As we shall see, the variation in Crinoids is not merely apparent, but is a question of pigment.

5. *Holothuria*.—Of the external colouring of the sea-cucumbers there is not much to say ; as a rule they tend to be dull and dark in tint. Black or brown colours are not uncommon, while greenish-violet or gray also occur. The relative dulness of the tints is curious in view of the fact that the pigments are numerous and often brilliant.

THE PIGMENTS OF ECHINODERMA

1. The pigments of the star-fishes, as their colour and their distribution indicate, are largely lipochromes, and are probably very similar in nature to those of the Crustacea. As in the Crustacea, the pigments are not confined to the skin but occur also in the internal organs and especially in the ovary. M'Munn (1883) describes hæmatoporphyrin, on spectroscopic grounds only, in the integument of *Asterias glacialis*. He has also shown that entero-chlorophyll occurs in considerable amount in the digestive cæca. As to the nature of the blue and violet pigments of star-fishes, there is much more difficulty. Krukenberg describes a blue pigment in *Astropecten auranticus* which is soluble in water, and is readily turned red by the action of heat, alcohol, and other reagents. The blue pigment does not apparently affect the colour of the organism during life, and the blue colour is only apparent when the superficial lipochrome has been removed by alcohol. Krukenberg regarded this blue pigment as being in all probability identical with cyanein, the blue pigment of jelly-fish. It does not seem to be the same as the blue pigments of Crustaceans.

2. As to the pigments of the Ophiuroids there is even less to be said ; they are probably for the most part lipochromes. Points of interest about both groups are first, as we have already noticed, the tendency for the colours to be more intense in deep-sea than in shallow-water forms, and second the greater instability of these bright pigments. There is reason to believe that this is due to the same cause as the similar phenomenon in the Crustacea, namely, that diminished power of secreting lime which is characteristic of abyssal organisms.

3. The pigments of the sea-urchins are more numerous and more difficult. Lipochromes probably occur, but apparently do not markedly predominate in the production of the surface coloration. The pigment enterochlorophyll is apparently common and present in considerable amount. Other pigments have been described, but are somewhat imperfectly known. As an example of the colour phenomena to be seen in the group, we may take the common *Echinus esculentus.* This is usually of a purplish colour with green spines tipped with violet. The perivisceral fluid and blood are a deep claret-colour, and apparently contain both " enterochlorophyll " and a brown pigment called echinochrome. If this claret-coloured fluid be exposed to the air it turns bright green — the colour of the spines. The change has been supposed to indicate the presence of a respiratory pigment, but there is as yet no certainty.

4. Among the Crinoids there is, as in the sea-urchins, a marked tendency for the lipochromes to be masked or replaced by more complex pigments.

Moseley discovered in several species of *Pentacrinus* a complex pigment, which he called pentacrinin. We have already mentioned the variations in colour which his specimens presented.   He found that the purple specimens yielded a pigment which formed a pink solution in acidified alcohol, the solution turning blue-green with ammonia.   Both solutions yielded banded spectra differing in the two cases.   Pale or yellowish-coloured specimens, on the other hand, yielded a green solution to alcohol, but the solution turned pink with acid, and apparently contained the same pigment as the purple specimens.   Moseley suggests that the reaction of these specimens was probably alkaline during life, and that the pigment was therefore present in its alkaline form.   Curiously enough Moseley obtained another specimen of *Pentacrinus* in which the complex pigment was entirely absent, and a lipochrome only was found.   It is, however, probable that the purple specimens contained a lipochrome mixed with the pentacrinin.

The species of *Antedon* show similar variations in the nature of their pigments.   In *Antedon (Comatula) rosacea*, Krukenberg describes yellow, red, and brown pigments, all nearly related to one another, and all soluble in water ; they do not give banded spectra. In another *Antedon* Moseley found a purple pigment which he calls *antedonin*.   It formed in alcohol when dilute a pink solution which turned orange with acid and violet with ammonia, and gave banded spectra differing in the different conditions.   These two peculiar pigments, antedonin and pentacrinin, resemble in several respects the pigments of the enterochlorophyll or chætopterin group, but have not been reinvestigated since Moseley's work. Moseley found antedonin also in a deep-sea Holothurian.

5. Among the Holothuria black or brown pigments are widely distributed, and may be identical with those occurring in the Crinoid *Holopus* and the dark-coloured Ophiuroids. In addition, numerous bright pigments occur, but are rarely important in the production of surface coloration. Among them lipochromes are probably common, especially in internal organs. In *Ocnius brunneus*, according to M'Munn (1889), the ovaries are pale blue, turning red and yielding a lipochrome when treated with alcohol and other reagents. This is an interesting case, for it seems to be the only one mentioned in the literature in which an identity with Crustacean pigment can be definitely maintained.

RELATIONS TO PIGMENTS OF CRUSTACEA

Looking at the pigments of Echinoderms as a whole, we must remark, first, the prevalence of lipochromes, especially in the star-fishes and brittle-stars where they seem to form the predominant pigments. Again, the case just mentioned suggests that the blue Crustacean pigments—the so-called lipochromogens—are found also in Echinoderms, though there is as yet no evidence to prove that they give rise to external colours.

So far, there is an analogy with the Crustacea, but we find further that the pigment called enterochlorophyll is common in the digestive cæca, in the perivisceral fluid and probably elsewhere in Echinoderms, while in Crustacea it is, at most, rare.

Associated probably with the presence of this pigment we find numerous bright-coloured integumentary pigments, which may be of the nature of derivatives, as well as complex pigments like antedonin and pentacrinin whose relations are more doubtful.

Another point of contrast with the Crustacea is the prevalence of dark pigments, which, though most characteristic of Holothurians, tend to appear more or less frequently in the other groups.

CHAPTER VII

THE COLOURS OF THE LEPIDOPTERA

Insect Coloration and its General Relation to Physiology—
Pigments of Caterpillars—Intrinsic and Derived Pigments
—Mr. Poulton's Experiment—Meaning of Derived Pig-
ments—Other Characters of the Coloration of Caterpillars
—Colours of Butterflies, Pigmental and Optical Colours
—The Pigments of the Pieridæ—Pigments of other
Butterflies—Origin of the Pigments of Butterflies—Con-
trast between the Pigments of Butterflies and those of
Caterpillars—Pigments and Mimicry.

THE colours of insects are in many cases so conspicuous that they have always attracted much
attention, and of late years few biological subjects
have been more keenly debated than the meaning
and uses of the tints and markings of caterpillars
and butterflies, bees and flies. The frequent similarity in colour between insects and their environment,
or between insects not nearly related, or on the
other hand the exceeding conspicuousness of some
well-protected insects, are facts which are obvious to
every one, and which have therefore attracted widespread interest. Until very recently, however, this
interest has been chiefly confined to the external

aspect of the colours, and even now little is known of their meaning to the organism in which they occur.

We do not propose here to consider in detail the characters of the colours of insects in their relations to the habits of the species. The subject has been most fully worked out for the Lepidoptera, and the facts of the case, as well as the conclusions drawn, will be found in Mr. Poulton's *Colours of Animals* and Mr. Wallace's *Darwinism*. Here we are concerned more with the proximate origin of colour than with its ultimate justification.

The colours of insects are of especial importance to the comparative physiologist on account of the general tendency of the group to exhibit a life-history divided into two sharply-contrasted stages: the larval stage in which growth and nutrition are at their maximum, and an adult stage in which these are almost at a standstill, while the activities are directed to the maintenance of the species.[1]

As might be expected the colours of the two stages are often very sharply contrasted. In the majority of cases the colours of the larval stage tend to be sober as compared with the often bright colours of the adult, just as generally speaking the larva may be called sedentary as compared with the active imago. Further, since the cuticle of the larva is usually little differentiated as compared with that of the adult, and we have already considered the relation existing between a differentiated cuticle and the

[1] For an exceedingly interesting discussion of this and other points connected with the physiology of insects, the reader should consult Dr. David Sharp's "Insects" in the *Cambridge Natural History*.

development of optical colours, it is almost unnecessary to add that optical colours are almost always absent from the larva, however brilliant they may be in the adult. These statements refer especially to the Lepidoptera where the contrast between the two parts of the life-history is very marked. It may also be noted that caterpillars are free living forms, more or less completely exposed to air and sunlight, and we know that, whatever be the explanation of the fact, this mode of life tends to favour the development of pigment. We are thus justified in selecting the Lepidoptera as a suitable group with which to begin the study of the colour phenomena of insects. Both stages of the life-history frequently display beautiful coloration, so that the contrasts between the pigments of larvæ and adults present themselves here in the most vivid form.

PIGMENTS OF CATERPILLARS

Butterflies and caterpillars are such familiar forms that we need not discuss their characters, either of structure or of coloration, but may pass at once to consider their pigments.

It is remarkable that in spite of the attention which has been directed to the colours of the Lepidoptera very little is certainly known of the chemical nature of the pigments of caterpillars, and this in spite of the fact that, as has been already mentioned, there are several painstaking researches on the colouring - matters of butterflies' wings. What information we have is mostly due to Prof. E. B. Poulton, who has experimented especially on the relations existing between larvæ and their environment. Mr. Poulton divides the colours of larvæ into two classes : (1) colours due to pigments derived from the food, and (2) colours due to pigments formed by the larvæ. Of the chemical nature of the pigments belonging to the second group we unfortunately know nothing ; they appear to be usually though not invariably dark in colour, and are deposited in the cuticle or in the epidermis. Their stability and their insolubility in alcohol suggest the possibility that they may belong to the same group as the dark pigments found in the adult.

The pigments belonging to the first group are green, yellow, or brown and are described by Mr. Poulton as " modified chlorophyll " and xanthophyll. Xanthophyll, as we have already explained, is the lipochrome pigment which can be obtained from the solutions of chlorophyll formed by steeping green leaves in alcohol. The chlorophyll was recognised in the larvæ solely by certain resemblances between the spectroscopic characters of the larval pigment and those of green leaves. These pigments occur primarily in the digestive tract, whence they seem to reach the blood ; while from the blood they may be deposited in the subcutaneous connective tissues. In most cases it is only when they occur in the last position that they are important in producing coloration. In his first paper (1885) Mr. Poulton supported his position that these pigments were derived from the food by a number of arguments, as well as by the results of spectroscopic examination. More recently (1893) he has been able to demonstrate experimentally that the pigments are absent in the larvæ when they do not occur in the food.

MR. POULTON'S EXPERIMENT

For the purpose of this experiment Mr. Poulton obtained a large number of the eggs of *Tryphæna pronuba*, the Yellow Underwing. Immediately after hatching, the larvæ were divided into three sets. The first set was furnished with the yellow etiolated leaves from the heart of a cabbage, the second with the white midribs from the same leaves after the removal of the whole of the blade, and the third with the ordinary green outer leaves of the cabbage. All the specimens were kept in the dark, in order to avoid the risk of the conversion of the etiolin into chlorophyll, and were only brought to the light to be examined and fed. The larvæ in the first and third sets developed well and showed almost identical colouring. In the early stages they were mostly pale green, but as development proceeded the colour deepened to dark green, and before maturity was reached most had turned dark brown. In addition to the conspicuous green or brown pigment in the connective tissue, or in the epidermis, the cuticle showed spots and patches of dark pigment, which were most conspicuous in the regions where the cuticle was especially thick, such as the head, the true legs, etc. This pigment is called by Mr. Poulton "true" pigment, and as is shown by the condition of the second set of larvæ is not dependent for its formation upon the presence of pigment in the food.

The larvæ in this second set developed badly and out of a large number only one reached complete maturity. This was probably due to the fact that the cut mid-ribs dried up rapidly and rendered it difficult for the· larvæ to obtain sufficient food, especially in the early stages when the mandibles were weak. The larvæ in this set were throughout of a pale cream colour with no trace of green tint, but with the cuticular markings quite distinct ; the cuticular pigment being as usual most conspicuous in the parts of the body where the cuticle was thickest. The single larva which attained maturity had, therefore, cuticular pigment developed as fully as usual, but was otherwise quite colourless.

From this experiment Mr. Poulton draws the natural conclusion that the green or brown ground-colour of these larvæ is produced by modified pigments derived from the food, and is dependent for its production upon the presence of these pigments in the food, while the dark cuticular pigment is not so dependent. Further, as the colour of the larvæ fed on yellow leaves is the same as that of larvæ fed on green leaves, it would seem that the larvæ can transform the yellow pigment into a green one. The exact nature of this green pigment is still uncertain ; it is unlikely that it is chlorophyll. The brown colour is probably the result of an oxidation of the green pigment. By developing control specimens in the light, Mr. Poulton found that the only difference produced by the absence of light was the assumption by the larvæ of a brown ground-colour before reaching maturity, while larvæ developed in light usually remained green till maturity.

Although there has been relatively little experimentation, there is, according to Mr. Poulton, considerable evidence to show that yellow, green, and brown ground-colours in vegetarian larvæ are all due to pigments derived from the food. The pigments frequently colour the blood very markedly ; it will be recollected that the blood in insects has probably no connection with respiration, but is hæmolymph, primarily concerned with nutrition. The blood of the pupæ and adults is often likewise green and coloured with these derived pigments, but as yet there seems to be no instance described in butterflies or moths in which these pigments assist in coloration (a possible exception will be discussed when we consider butterflies). They are retained within the body, however, and not infrequently colour the eggs and thus the newly-hatched larvæ. The break between larval and adult coloration seems in Lepidoptera to be complete. But lipochrome pigments at least are not always absent in adult insects, for Zopf describes red and yellow pigments belonging to this group in several small leaf-eating beetles. In *Lina populi*, for example, which lives on the leaves of the poplar, a red lipochrome colours the wing-covers and part of the abdomen, it is found in the secretion of the salivary (?) glands, and occurs associated with oil-drops in the ova. One is tempted to suppose that its presence here is associated with the prolongation of the larval diet into adult life.

MEANING OF DERIVED PIGMENTS

We must now proceed to consider what the abundant occurrence of derived pigments in insects may mean in the physiology of the individual. Perhaps the following may not be thought to transgress the bounds of legitimate speculation. It is well known that the fats of different animals are different both in their physical and chemical characters. Now it has been found by experiment that if an animal A be fed to excess with the fat of another animal B, and the body of A be subjected to subsequent examination, then the fat deposited in it will not exhibit the normal characters of the fat peculiar to the animal, but will partake more or less of the characters of the fat of the food. In other words, an animal is unable to impress its own individuality on the fat of its food, if this be ingested in excessively large quantity. Now we have seen that in all probability the derived pigments consist, at least in great part, of lipochromes, and we know that in most cases the lipochromes tend to occur in association with fats ; in chlorophyll the presence of fat has indeed been directly affirmed. Is it not possible that in the caterpillar—a notably voracious feeder—a process occurs quite similar to that described above for mammals ? That is, may not caterpillars, which have a practically unlimited food-supply, be unable to completely assimilate all the fat ingested, but yet have the power of storing up in their tissues this extra fat, and with it the pigment with which it is associated in the food ? The process would thus be very similar to that which occurs in the salmon, where, as we have seen, both fat and pigment are transferred from the muscles to the ovaries during the period of fasting. If the assertion that in the salmon the pigment is derived from the food is correct, the two processes would be very closely analogous. According to this theory, food containing lipochrome pigment is richer diet than food without it, and leads to the deposition of extra reserves in the tissues and indirectly to additional pigmentation. The facts observed by Mr. Poulton that the pigments are frequently found associated with fat, and tend to occur in the connective tissues, seem to support this suggestion. It is also not inconsistent with the observed fact that the pigments remain within the body of the moth, and may be used in the formation of the eggs and so passed on to the second generation. As nutrition in the butterfly is unimportant, there is reason to believe that the caterpillar must provide the nutritive substance subsequently employed in the formation of the yolk. Yolk is usually remarkable for the considerable amount of fat which it contains, and the lipochrome may be simply transferred with the fat.

OTHER CHARACTERS OF THE COLORATION OF CATERPILLARS

As to the other characters of larval coloration, Mr. Poulton describes cases in which colourless bands of fat give the appearance externally of white stripes. It is also of interest to note that the blood of the larvæ, when exposed to air, oxidises and becomes dark-coloured or black. It is possible that some such oxidation process accounts for the deepening in tint of the true or cuticular pigment when the larvæ are exposed to the air.

We do not propose here to consider the question

of variable coloration, or the nature of the adjustments by which it is rendered possible. Those who are interested in the question will find it discussed in numerous other popular books besides those already mentioned. It is, however, of some interest for our purpose to note that, according to recent research, it is not so much the colour of the environment which directly affects the larvæ as the intensity of the light (see Garbowski). In other respects also some of the first crude statements made on the subject are being modified. It may, indeed, be accepted as a general truth in biology that whenever a series of phenomena seem to be susceptible of an extremely simple and beautiful explanation, that explanation is wrong and founded on an imperfect acquaintance with the phenomena in question. All recent progress in biological theory has shown that life is not simple but complex, and has involved a substitution of extremely complex theories for simple ones.

THE COLOURS OF BUTTERFLIES

In butterflies the colour-phenomena are much more complex than in caterpillars, for while in the latter the colour is a direct pigmental effect, in the former the two factors of structure and pigment occur in combination. In both kinds of colour in butterflies the important elements are the scales of the wings. These are outgrowths of the chitinous cuticle, and consist of a double membrane ; the outer membrane is frequently much differentiated, and may display rows of blunt projections, which are

thought to be of importance in colour-production.
The two membranes are connected by bridges of
chitin, and pigment granules may be deposited both
in these bridges and in the outer membrane.

Having regard to the small size of the scales, it
will be readily understood that there is frequently
great difficulty in determining whether a particular
colour is due to pigment or structure ; the fact that
a scale showing structural colour frequently contains
pigment in addition further complicates the matter.
We thus find that there is much difference of opinion
among observers as to the cause of particular colours.
This occurs especially in the case of *dichroic* scales,
those displaying one colour by transmitted light and
the complementary colour by reflected light. Some
regard this effect as purely structural, others main-
tain that the colours are due to dichroic pigments.
Making due allowance for difficulties of observation,
it seems, however, certain that blue in butterflies is
always a structural colour, while green, black, and
white are at least usually due to structure. Other
of the structural colours are readily recognised by
their metallic brilliancy and changing glow, which
give to the butterflies possessing them an appearance of surpassing beauty. Many will recall Mr.
Wallace's description of how his heart beat fast
and his brain reeled when his perseverance was
rewarded, and he captured with his own hands
one of the finest of these living gems in the Malay
Archipelago.

In connection with these colours Urech (1893)
notices one little point of some interest. He found
that in some cases the scales display under the

microscope a wonderful play of colours which is quite invisible to the naked eye. He suggests reasonably enough that it is in no respect improbable that these colours may be visible to other insects although not to us except by the aid of optical instruments, and that therefore they may quite alter the appearance of the insects as seen by other butterflies. Such observations are of interest as tending to exercise a check upon the purely subjective treatment of questions of colour-resemblance.

THE PIGMENTS OF THE PIERIDÆ

As an introduction to the discussion of the pigments of butterflies we may take Mr. F. Gowland Hopkins's laboriously patient work on *The Pigments of the Pieridæ.*

The Pieridæ include a large number of butterflies, among which the most familiar are our common Cabbage Butterflies or Garden Whites; they are very widely distributed, and are said to exhibit in a striking manner the phenomena of mimicry. In America especially they are said to mimic very closely the (by hypothesis) well-protected Heliconidæ. The Pieridæ exhibit a relatively simple plan of coloration; optical colours are absent or little developed; and we shall see later that this greatly limits the possible colouring. In fact we find that the colours are mainly white, yellow, and black. The yellow, in accordance with a rule which is exceedingly prevalent among butterflies, may deepen into orange or red, this occurring most frequently in species from warm climates. The Heliconidæ exhibit similar types of colour-schemes.

In the Pieridæ the pigments, according to Hopkins (1896), occur in three ways—(*a*) first uniformly distributed through the chitin, apparently of the outer layer of the scale : the black and brown pigments occur in this position ; (*b*) in granules between the two layers of scales, according to Hopkins—that is, probably in the bridges of chitin already described : white, yellow, and red pigments occur in this position ; (*c*) between the two chitinous lamellæ of the wing : a green pigment is described in this position.

The black and brown pigments, as elsewhere in butterflies, are characterised chiefly by their insolubility and their stability ; they are not attacked by any agents which do not attack chitin. They are possibly the same as the dark cuticular pigments of the larvæ, and may have some connection with the substances which occur in the blood and oxidise rapidly when it is exposed to the air. In the Pieridæ they are important in directly giving rise to the dark patches and markings on the wing, but in many butterflies they probably aid in the production of structural colours.

The white pigment of the Pieridæ is, as we have seen, uric acid itself. It can be extracted from the tissues by dilute alkaline solutions, which yield a copious precipitate on acidification. The precipitate, if treated with nitric acid and heated, leaves a residue which turns purple with ammonia — the murexide reaction, so characteristic of the uric-acid group of substances.

The yellow or orange pigment is readily soluble

in hot water, forming a bright yellow solution with a green fluorescence. The pigment is insoluble in organic solvents, is acid in reaction, precipitates on cooling the watery solution, and gives the murexide reaction. It thus belongs to the uric-acid group, and has been called lepidotic acid by Hopkins. As already stated, it occurs as one of the ordinary waste products of the organism, at the time of the emergence from the pupa. It is a derivative of uric acid, and can be artificially produced by heating uric acid with water in sealed tubes to a high temperature. It is somewhat interesting to note that by heating the yellow pigment with dilute acids in a water-bath, Hopkins obtained a purple colouring-matter, yielding a spectrum with two bands. This pigment does not, however, appear to occur naturally, unless the purple tips of the fore-wings of *Anthocharis ione* be due to it.

Although uric and lepidotic acids frequently occur in the same wing, Mr. Hopkins believes they never occur in the same scale; pale yellow patches are due to a small amount of pigment in the scales and not to an intermixture of the two acids. Orange patches are due to a large amount of the yellow pigment, but red ones are due to the presence of a red pigment. They yield their pigment to hot water, but the aqueous solution is yellow. On evaporation this solution yields, however, a red residue. The residue gives the murexide reaction, and also yields the purple substance, lepidoporphyrin, just as the yellow does. Mr. Hopkins (1892) is of opinion that the difference between the red and yellow is either purely physical or due to the association of the yellow acid body with a base.

The only other pigment found in the Pieridæ is the green one described by Mr. Hopkins as occurring between the laminæ of the wing. This is soluble in cold water, contains iron, and gives a spectrum with a single well-marked band in the red; it is perhaps a blood pigment, and is common among the Pieridæ. It, however, rarely affects the surface coloration, except in the male of *Nepheronica lutescens*, where the blue ground-colour is said to be due to this pigment shining through the colourless scales.

In general, therefore, the pigments of the Pieridæ are few in number, and the colours are usually due to simple pigmental effects. The pigments are uric acid, lepidotic acid and that modification of it which forms the red pigment, the dull black and brown pigments, and the rare green pigment. These pigments rarely occur intermixed, and structural colours are typically absent.

PIGMENTS OF OTHER BUTTERFLIES

With regard to the pigments of other butterflies there is much less certainty. Hopkins has not succeeded in obtaining the murexide reaction outside the limits of the Pieridæ, and is inclined to think that the pigments of the uric-acid group are confined to this family. Urech, on the other hand, assumes that the pigments of butterflies in general are modified waste products, which Mr. Hopkins also admits as possible. The fact that almost all the bright-coloured pigments are soluble in hot water, as well as some other reactions, seems to support this view.

Among butterflies in general the following colours

are wholly or in part due to pigments deposited in the tissues :—white (in part), yellow, red (at least in part), very rarely green, brown, black (in part). White as a pigmental colour is probably rare outside the Pieridæ ; it often exhibits a changing iridescence like that of mother-of-pearl, and is then structural.

In the Marbled White (*Arge galathea*), an insect with a simple colour-scheme in black and yellowish white, a white pigment occurs which turns bright yellow with alkalies. Of yellow pigments in butterflies not included in the Pieridæ very little is known ; none give the murexide reaction. Many red pigments, on the other hand, give a reaction the same as that displayed by the red pigment of the Pieridæ. That is to say, it has been noticed by several observers that in many cases scales of a bright red or scarlet colour yield a *yellow* solution when treated with hot water, but if the solution be evaporated to dryness a *red* residue remains. Similarly, the red colour is turned yellow by the application of acid, but the red colour may be restored by the addition of ammonia. This is the so-called "reversion effect" of Mr. Perry Coste, and strongly recalls the relation described by Mr. Hopkins as existing between the red and yellow pigments of the Pieridæ. Red pigment showing this peculiar character occurs, for example, in *Deilephila elpenor*. The phenomenon is of common though not universal occurrence, some reds being quite unchanged by acids.

Green in butterflies presents many difficulties. It may be entirely structural, and arise by surface markings or by the superposition of scales, as in the species of *Nematois* (Spuler). Again, from the

green scales of *Papilio eurymedes*, Urech extracted a
yellow pigment which was almost insoluble in water,
but which dissolved readily in hydrochloric acid.
He adds as a note, however, that the scales retained
their green colour after treatment with acid and
ammonia. It is almost impossible to doubt that in
this case the green colour is structural, the part
played by the yellow pigment being uncertain.
Further, he found that the green scales of *Thecla
rubi* are yellow by transmitted light, and almost
colourless when the light falls from the base of the
scale upwards, while to hydrochloric acid they yield
a yellow pigment. These two cases seem to suggest
that in butterflies, as in birds, green may be produced by a combination of a yellow pigment and
a structural modification. On the other hand, from
the green scales of *Sphinx nerei* Urech extracted a
pigment which was slightly soluble in water and
readily soluble in acid and ammonia. Of the three
solutions the first was greenish-yellow, the second
orange-yellow, and the third green. The addition of
ammonia to pigment turned yellow by acid restored
the green colour. This fact would suggest that the
green pigment of *Sphinx nerei* is derived from a
yellow, in much the same way as is the red pigment
of *Deilephila*. These facts suggest the possibility
that a yellow acid body in some way related to
lepidotic acid occurs in butterflies outside the Pieridæ,
capable of acting itself as a pigment and also of
giving rise to red (or green ?) pigments.

Urech (1890) has made numerous observations on
the relation between the colour of the urine in butterflies and the prevalent colours of the scales. His

table seems on the whole to confirm the idea that a relation exists between the two, and therefore that the pigments are waste products, but observations on colour alone are perhaps not of very great value.

The dark pigments, as we have already seen, are probably identical in all butterflies; a pure black colour is, however, in most cases due to the sculpturing of the surface.

In general, although the evidence is still incomplete, we seem justified in believing that the pigments of butterflies are either modified waste products or the dark pigments. From the scales of a large number of butterflies Urech did not succeed in a single instance in obtaining any pigments which were soluble in alcohol, benzol, or the other common organic solvents. That is, if we may trust his results, lipochromes are absent in the adult. This is interesting, not only because of the prominence of these pigments in the larvæ, but because they are so important in coloration in animals nearly related to the insects. In Crustaceans, for example, as we have seen, the lipochromes colour the shell, the skin, and the ova. We have Mr. Poulton's authority for saying that even in Lepidoptera lipochromes still colour the ova, though here they are probably derived pigments and no longer native, but they have in the adult entirely disappeared from the cuticle and from the skin so far as external coloration goes, being replaced by modified waste products. As a probable exception we may notice the green pigment found by Hopkins in the Pieridæ; this must surely be one of Mr. Poulton's derived pigments!

ORIGIN OF PIGMENTS OF BUTTERFLIES

Let us now consider the meaning of this utilisation of waste products, and first of all their immediate origin. Urech suggests that they arise from the breaking down of the substance known as nuclein, which has recently been rising in importance in the eyes of physiologists. Nuclein is a complex substance forming a considerable part of the nuclei of all living cells ; it is derived from albumen, from which it chiefly differs in containing a large amount of phosphorus. According to Urech the nuclein of the leucocytes in butterflies undergoes a process of breaking down, yielding the nuclein bases, that is, such waste products as uric acid, guanin, adenin, hypoxanthin, etc., and also phosphoric acid and albumen. The nuclein bases then become further modified into the pigments. The different colours of the pigments Urech describes as due to an increase in the molecular weight of the substances, the yellow pigment being thus simpler chemically than the red one.

Mr. A. G. Mayer gives a somewhat different account of the development of the pigment of the scales in butterflies. We may consider first his account of the development of the scales themselves, which is interesting in several respects, and is illustrated by some very good figures. These show clearly that the scales develop from prolongations of certain formative cells of the epidermis, which project outwards beyond the level of the other cells. The scales at first contain protoplasm, which is continuous with the protoplasm of the formative cell, but as development proceeds the protoplasm is withdrawn, leaving the scale empty ; the formative cell also undergoes a process of degeneration. The scale itself, of course, consists of chitin formed apparently at the expense of the original protoplasm.

Mr. Mayer's account of the formation of the pigments is as follows :—After the protoplasm is withdrawn from the scale, this for a time is entirely empty and contains nothing but air ; a little later the hæmolymph of the pupa enters it and gives it an ochre-yellow colour. The fresh hæmolymph is of an amber colour, but when shed it speedily turns a turbid ochre-yellow colour, the same colour as that manifested by the lymph of the scales. After remaining for about twenty-four hours in the ochre-yellow stage, the scale begins to acquire gradually the characteristic colours. Mayer is of opinion that this development proves that the pigments are formed from the degenerating hæmolymph contained in the scale. He endeavours to prove his position by recounting some observations on the reactions of the hæmolymph to chemical agents. Thus " warm concentrated nitric acid " gives a " chrome-yellow " colour changing to " reddish-orange " with ammonia, the colour being very like that of a pigment-band on the wing of the moth experimented with (*Samia cecropia*). In what respect, however, the colour differs from that always given by proteids when treated with these reagents—which of course constitute the xanthoproteic reaction—is not mentioned. Although the reagents employed by Mr. Mayer were all too powerful for us to lay any stress on the chemical side of his work, yet on the histological side there

are several points of great interest.   Thus he notices
that in the case of the larger scales it is not infrequent for leucocytes to enter the cavity of the scale
and there disintegrate.   These leucocytes apparently
originate from the cells of the fatty body, a structure
which has probably much to do with excretion in
insects.   This passage of " wandering cells " outward
to be deposited in the superficial structures is a
phenomenon which we shall frequently have to
mention in connection with pigmentation.   Not
infrequently, as in the leech, the cells are themselves
pigmented, and deposit their pigment in the skin.
It would be a fact of some interest if Urech's
hypothesis is correct, and these cells by their
disintegration be found to give rise to the peculiar
pigments of butterflies.   It is of course quite possible
that Mayer is also right, and that the hæmolymph
does give rise to certain of the pigments, especially
those of relatively dull tints.   As, however, it is
the bright-coloured waste products which are most
important in butterflies, we must return to our
consideration of these.   Urech's suggestion is of
much interest so far as it goes, but we require to
know why this decomposition of nuclein should go
on in the adult and not in the larva.   In attempting
to answer this question we must consider the differences between larva and adult, and the physiology
of the state which lies between the two.

CONTRAST BETWEEN PIGMENTS OF BUTTERFLIES AND CATERPILLARS

We have already considered some of the characters of the caterpillar—how it eats far in excess of its immediate requirements, how it is relatively sluggish, and remarkable for its rapid growth. There comes a period, however, when the caterpillar ceases to eat, ceases to grow, becomes absolutely sedentary, and passes into the pupa stage. Here within its pupal covering fundamental changes take place ; there is first an extraordinarily thoroughgoing destruction of tissue and then an equally thoroughgoing reconstruction. Now if we turn to higher forms whose physiology is well known, we find that such extensive changes in tissues are associated with a large and increased production of nitrogenous waste products. It is natural to conclude that this also occurs in the pupa. The butterfly is built up by the aid of the reserves stored by the caterpillar, but in the course of the reconstruction there must be an enormous production of waste substances. Unfortunately we know relatively little of the physiology of insects, but there is at least one interesting research by M. L. Cuénot. This naturalist worked at the Orthoptera and found that besides the Malpighian tubes, the familiar excretory organs of insects, certain cells of the fatty body stored up waste products. In the Orthoptera — cockroaches, locusts, grasshoppers, etc.—waste products are apparently not employed in coloration ; and Cuénot found that throughout

life the fatty bodies contained deposits of urates—
that is, salts of uric acid—which were apparently
never eliminated, not even at the moment of
emergence from the pupa.   In the Lepidoptera, on
the other hand, which are more perfectly organised
animals, we find that these useless waste products
are perhaps in part eliminated at the time of the
emergence from the pupa state, and in part utilised
as pigments.   The wings are non-vital parts of the
body, at a distance from the essential organs ; we
can therefore readily understand that associated
with the advance in structure from the Orthoptera
to the butterflies we should have a removal of
waste products from the fatty body, to be in part
eliminated, in part stored up in the wings.   Further,
if in vertebrates some of the most important advances
are associated with the development of a more efficient
kidney, if the steps from fish to snake and from
snake to bird are accompanied by a diminished
deposition of waste products in the skin, is it not
possible that a similar process occurs in butterflies ?
That is, may not the occurrence of a large amount of
bright pigment in the wings of a butterfly indicate
a relatively low degree of specialisation ?   We have
seen that in the Pieridæ structural colours are
practically absent, while pigmental colours are
vivid and striking.   When, however, we pass to
such families as the Lycænidæ and the Apaturidæ,
we find that pigmental colours decrease in importance, and structural colours are responsible
for the major part of the effect.   Is it not possible
that this marks a genuine advance ?

PIGMENTS AND MIMICRY

Mr. Hopkins notes in relation to the question of mimicry that the pigments of the Heliconidæ are not the same as those of the Pieridæ, for though soluble in water they do not give the murexide reaction. He regards this as evidence against the views of those who hold that the resemblances of mimicry may be due to relationship between the forms. We do not propose to discuss the question of mimicry here, but may just note that mimicked species usually display what are known as " warning colours," which apparently, in butterflies at least, are always pigmental colours, or the simple types of optical colour like white and black. Now as there are in butterflies, apart from structural colours, only about three pigments to choose from (yellow, red, brown, and rarely green), the range, if we may so speak, is somewhat limited. Apart from the question whether the presence of the yellow and red pigments in considerable amount is not a sign of little specialisation, the frequent absence of structural colour from these " warning " forms is at least probably such a sign. Instead therefore of supposing that the Heliconidæ have, in Mr. Wallace's words, " acquired lazy habits " and a slow flight because they are uneatable, and the Pieridæ because they resemble the Heliconidæ, may we not rather suppose that the slow flight and " warning " colours in both cases are due to the same cause, the relatively low organisation which renders pigmentation by waste products possible, which makes brilliant

M

optical colours impossible? The resemblance between "mimicking" and "mimicked" forms is, of course, not wholly explained by the small number of pigmental colours which can appear in a butterfly devoid of optical colours, for the resemblance is often as much in markings as in actual colour ; but if we deny the "mimicry," we must, apparently, fall back upon those "laws of growth" which many, perhaps justly, find so unsatisfactory.

CHAPTER VIII

THE COLOURS OF INSECTS IN GENERAL AND OF SPIDERS

Optical and Pigmental Colours in Insects—Colours of the different Orders of Insects—Colours and Pigments of Orthoptera—Optical Colours of Neuroptera—Coloration of Beetles — Pigments of Hemiptera — Colour Resemblances between Diptera and Hymenoptera—Contrast between Coloration of Lepidoptera and other Insects— General Aspects of Insect Coloration — Variation in Colour : (1) Natural, (2) Artificially produced — The Colours of Spiders : (1) Optical, (2) Pigmental—Development of Colour—Variation in Colour—Colours of the Sexes—Markings of Spiders.

AMONG insects other than butterflies the colours are often striking and beautiful, and are again divisible into pigmental and optical. Probably, as a rule, when an individual exhibits brilliant colouring, this is mainly due either to pigment or to structure, the two types of colouring being rarely both conspicuous at the same time. But this is true only of the bright pigments, the dark being common in insects displaying much structural colour. This is due to the fact that the dark pigments are present in excess only when the cuticle is well differentiated, and this is also the condition necessary in the general case for the development of optical colour. Mr. Poulton speaks of larval Lepidoptera exhibiting the dark cuticular pigments in the places where the cuticle is thickest *because* the epidermal pigments would be useless in these spots, the cuticle being too thick to allow them to shine through ; a comparative survey, however, makes it more probable that there is a direct association between the pigment and the thickened cuticle. An excellent example is afforded by the dark colours of many beetles with their thickened elytra, which form a marked contrast to the browns or greens of the Orthoptera with their relatively little developed cuticle.

Another important point about the colours of insects is the relation of food to colour. Mr. Poulton's experiments on caterpillars have shown that the pigments of the food may find their way to the blood or to the tissues, and owing to the thinness of the cuticle be instrumental in producing the typical coloration. The prevalence of green colours among herbivorous forms with little developed cuticle, *e.g.* the Aphides or plant-lice, certainly suggests that this also occurs elsewhere. The question has been considerably confused by the common habit of calling such derived pigments chlorophyll. In the Crustacea green pigments occur, *e.g.* in *Virbius*, which there is little doubt are merely combinations of lipochromes with other substances, perhaps bases, and there seems no reason why similar combinations should not occur in insects, with the aid of the derived pigments. At the same time there is no apparent reason why insects should not themselves produce lipochromes, and why such lipochromes should not occur in the cuticle as in the Crustacea. We have seen that lipochromes occur in the elytra of *Lina populi ;* according to Krukenberg they perhaps also colour the red lady-birds (Coccionellæ) and some other red beetles, but whether these are intrinsic or derived pigments remains undecided.

COLOURS OF THE DIFFERENT ORDERS OF INSECTS

It is interesting to observe the types of colouring prevalent in the more important orders of insects. Thus in the ORTHOPTERA optical colours seem to be practically absent, and deep black is also rare; shades of brown, green, and red are perhaps the commonest colours, as notice the brown "stick" insects, the brown and green (or sometimes red) grasshoppers and locusts, and so on. Associated with these peculiarities of colouring we notice the fact that the cuticle is in most cases but slightly developed. The relatively sober tints do not, however, prevent the Orthoptera often displaying great beauty of colouring. Thus Dr. D. Sharp describes the female of *Pneumora scutellaris* as being "one of the most remarkably coloured of insects." " She is of a gay green, with pearly - white marks, each of which is surrounded by an edging of magenta ; . . . the face has magenta patches and a large number of tiny pearly-white tubercles, each of which when placed on a green part is surrounded by a little ring of mauve colour." The male is much plainer, being

" of a modest, almost unadorned green colour " (*Insects*, p. 302).

It is impossible to speak of the colours of the Orthoptera without mentioning the remarkable leaf-insects (*Phyllium*). As is well known these insects exhibit an extraordinary resemblance to leaves both as to structure and colour. They are, however, when hatched not green but red ; after a few days, during which they feed greedily on leaves, the colour becomes yellow, and after the first moult it is greenish. After this the green becomes more and more intense after every moult (Becquerel et Brongniart), but it is asserted that when dying the colour changes, exhibiting the hues of a fading leaf. In dried insects the colour is very fugitive. All these characters suggest an origin from the pigment of the food, and MM. Becquerel and Brongniart have sought to prove this spectroscopically. They found that the pigment occurs in amorphous granules in the subcutaneous connective tissues, and that it gives a spectrum closely resembling that of the green leaves upon which the insects feed. As, however, the whole insect was employed for spectroscopic purposes, without apparently any effort being made to remove the gut or its contents, the observations are not very conclusive. The frequency in the order of green colours associated with herbivorous habit may suggest that derived pigments are common and important in coloration, but, on the other hand, we have cases like that of the species of *Mantis* where a green colour is not infrequent, and yet all are purely carnivorous. There seems also nothing intrinsically improbable in the idea that the green colour may be produced by a yellow pigment in combination with some simple effect of structure, and that progressive colour-change, like that of *Phyllium*, may have more relation to structure, *e.g.* thickened cuticle, than to pigment.

Besides the uniform brown and green colours in Orthoptera there is not infrequently a development of beautiful markings and patterns, especially in browns and yellows.   Dr. Sharp mentions as a very peculiar case that of *Corydia petiveriana*, where the wing-covers are spotted, but not symmetrically. In the resting position, however, the two overlap in such a way that the markings then appear completely symmetrical—a case of some theoretical interest.

On the whole there is reason to believe that the colours of the Orthoptera are to a considerable extent due to lipochrome pigments, whether intrinsic or derived is still uncertain.

In the NEUROPTERA, as exemplified by the dragon-flies and may-flies, the colours are in large part optical.   The gauzy wings frequently display beautifully changing tints, but in other cases instead of being transparent they show graceful patterns and markings in shades of brown (*e.g.* species of *Myrmeleon*).   The body also often displays bright optical colours, especially shades of blue.   These colours are associated with the development of a considerable amount of dark pigment, and are mostly very fugitive. Bright coloured pigments are relatively rare, and in general terms the colours are due to the dark pigments, or are optical colours produced by the differentiation of the cuticle.   The optical colours may be associated with a thickening and pigmentation of the cuticle as in the body, or may be simple inter-ference colours like those of the wings.

In the large order of beetles (COLEOPTERA) many varieties of coloration are displayed, some species being notably dull while others exhibit a brilliance rivalling that of the Lepidoptera. Black and deep brown pigments are very common, especially in the elytra or wing-covers, which are of course the con-spicuous features in most beetles. We have already noticed the relation between the thickened cuticle in this region and the pigment. The colour may be simply a dull, dead black as in many Carabidæ, or it may be a vivid metallic green or blue as in the rose beetle ; between the two there are many intermediate stages ; but the metallic tints, which are of course optical, are very common. Then there may be patterns and markings in brown and black, as in the very beautiful *Macropus longimanus* which is elaborately marked with shades of pink and gray. Finally, we find the development of bright red, orange, and yellow (lipochrome, Krukenberg) pigments, as in species of *Coccinella* and *Chrysomela*, producing patterns and markings when blended with black and brown. As a whole, however, probably optical colours are most common in beetles. A consider-able number of beetles are furnished with scales, but these are of less importance in coloration than in the Lepidoptera.

Among the pigments of beetles few have been investigated, but as to the minute details of colouring there are some facts of interest. Thus in *Luciola italica*, the Italian glow-insect, as described by Emery, the male has the prothorax coloured a clear red, while that of the female is paler and more yellow. The cuticle is, however, similarly coloured in both sexes, and the difference is due to the different colouring of the fatty bodies in the two. In both these contain concretions of urates, but in the female they are chalk-white, in the male a delicate rose colour. In the related *Lampyris*, on the other hand, both sexes have rose-coloured concretions. In *Luciola* the testes are also coloured red by concretions deposited in the connective tissue. The association between the urates and pigment is interesting, but Emery says nothing as to the characters of the pigment. In *Luciola* the adult does not eat, reproduction being the only object of existence at this stage. It is a point of some interest that we have here an indirect utilisation of the pigment of the fatty body in coloration, a condition which suggests a transition towards the complete utilisation of coloured excretory products seen in some of the Lepidoptera.

The HEMIPTERA display great variety in their coloration, and apparently contain many peculiar pigments. Optical colours are not very common but are far from being unknown ; thus *Eurymela* shows a faint metallic sheen, and *Libyssa signata* is marked by alternate bands of vivid metallic green and black.

Among pigmental colours we have not only uniformity of colour but also frequently beautiful patterns and markings. Our own familiar Bishops' Mitres (*Tropicoris*) exhibit considerable beauty of marking, while many of the large Indian and American forms have most beautifully marked and spotted wings. In these cases the fore-wings are not infrequently different from the hind, and the type of coloration suggests that seen in butterflies.

Among the pure pigmental colours brown, red, yellow, and green are common. A brilliant red colour is far from uncommon, and is displayed for example on the surface of the abdomen in *Nepa rubra*.

Among the pigments of the Hemiptera we must assign the first place to the carmine of *Coccus cacti*, not on account of its importance in coloration but because its commercial value has caused it to be somewhat fully investigated.

Carminic acid occurs in nature in the Coccidæ, especially *Coccus cacti*, and, according to Sorby, also in various species of *Aphis*. It is a glucoside, yielding a sugar when treated with dilute acid.

In the case of *Coccus cacti* the pigment occurs in large amount in the female (26 to 50 per cent of the body weight according to Krukenberg) and in less amount in the male ; to the dried insects at least it gives a dull red-gray colour rather than a pure full red. According to Mayer it occurs in drops near the periphery of the cells of the fatty body, the drops being less numerous in the case of the male. It occurs also in the yolks of the eggs and in the diffuse fatty body in new-hatched larvæ. Mayer says expressly that the pigment does not occur in the gut, but in another place he states that it markedly colours the fæces ; the anomaly is, however, nowhere explained. It is possible that the meaning is that the pigment is not found in the anterior part of the gut, but is introduced into it by the Malpighian tubules.

As to function, Mayer has no suggestion to offer, but is of opinion that the pigment is undoubtedly intrinsic and not derived. Krukenberg believed that it functions as a reserve, basing his opinion especially on the ground that the large amount present in the female is inexplicable on any other hypothesis, and that it is of glucoside nature. It may be pointed out, however, that chitin itself is a derivative of a carbohydrate, and often occurs in considerable quantity, and yet no one has suggested that it is a reserve product. It seems as yet impossible to decide the question of the function of carmine, but we may note that as we have already seen in the case of *Luciola*, the association of pigment with the cells of the fatty body is not unknown among other insects. Mayer describes colourless, crystallised bodies in the cells containing carmine, but does not discuss their nature ; they may of course again be urates. If so the contrast between the pigmentation of the sexes in *Luciola* and *Coccus* is striking in the extreme.

As to the other pigments of the Hemiptera, Sorby investigated those of Aphides, and described a pigment found in them as respiratory, but, according to Krukenberg, his results were due to an admixture of carmine and lipochromes. The remaining pigments do not appear to have been much investigated ; the characters of the bright red one found in various forms would be of great interest in a comparative study of insect pigments. The red pigment of *Pyrrhocoris apterus* is, according to C. Phisalix, closely related to carotin, and therefore a lipochrome.

Among other orders of insects the HYMENOPTERA

and DIPTERA may be mentioned as exhibiting a striking parallelism of coloration. In various instances (e.g. *Volucella, Eristalis,* etc.) this has of course been described as protective mimicry on the part of the flies, but it is difficult to look at collections of the two orders side by side without being struck by the similarities of colouring. In both cases the body is frequently covered with hairs which are important factors in coloration, but the hair may be completely absent and the surface may be then brilliantly metallic or black and polished-looking. Among pigmental colours in both cases black, dull brown, and yellow are the commonest. There seem to be no investigations on the pigments in either case.

CONTRAST BETWEEN COLORATION OF LEPI-DOPTERA AND OTHER INSECTS

In comparing the colours of other insects with those of the Lepidoptera, we have to notice that in the former the occurrence of much beauty or special-isation of colour in the larvæ is relatively rare. The larvæ not infrequently are of the form popularly known as maggots or grubs, and they are usually colourless or almost so. Similarly, outside the Lepidoptera, it is not very common to find the nutritive and reproductive stages sharply contrasted in large groups, many of the adults being not in-frequently both. Directly associated with this fact, or as the indirect result of the diminished specialisa-tion which makes food-taking possible to the adults, we find that the adults may display that type of coloration which is larval for the Lepidoptera, that is

especially coloration by lipochromes.   When, however, the adults, as in the beetles, are remarkable for
the great development of the cuticle, the coloration
may be due either to the presence of a large amount
of dark pigment or to structural colours.   Coloration
by waste products has not yet been described outside
the Lepidoptera, at least in detail.

General Aspects of Insect Coloration

It is impossible to conclude this section on the
colours of insects without touching, be it never so
slightly, on some of the interesting questions which
cluster around the subject.   As most of the current
theories of colour have been in large part founded on
insects, it is not surprising that the literature of the
subject has already reached enormous dimensions, so
that to attempt to abstract even the more important
papers would be impossible.   We have therefore
confined our attention very closely to the physiological side, and the result is of course professedly
incomplete.   Many of the omitted subjects, such as
the colours of the sexes, the relation between food
and colour, variable coloration, and so on, are, however, already in their rough outlines familiar to most
people, and their physiological side has been so little
investigated, that little more can be profitably said.
There are, however, one or two points which, on
account of their general interest, are worthy of more
extended treatment.

VARIATION IN COLOUR

1. *Natural.*—With regard to those variations of colour in insects, especially butterflies, which occur in nature there are some interesting observations. Much of their interest is of course due to the fact that butterflies show such a marked tendency to develop into geographical varieties distinguished primarily by their colour. Mr. Bateson, in his *Materials for the Study of Variation*, pp. 44, 45, describes several cases having especial reference to yellow, orange, and red pigments. Thus *Colias hyale*, the Pale Clouded Yellow (Pieridæ), is usually sulphur-coloured but may occur in white varieties, or may be of a "rich sulphur colour" with apical marginal patches of a red colour. A more interesting case is that of *Gonepteryx rhamni*, a British Pierid, in which the male is of sulphur-yellow colour, while a South European species, *G. cleopatra*, is very similar but has more orange on the fore-wings. Not only have forms closely resembling *G. cleopatra* been found in Britain, but intermediate forms have also been found, and also forms in which the fore-wings were broadly suffused with scarlet instead of orange. These facts are of interest not only on account of the relation which we have already noticed as existing between yellow, orange, and red pigments, but also on account of their importance in regard to questions of mimicry and of the effect of environment on colour. We shall see in the chapter on mimicry that such colour variations often occur simultaneously in very different genera, and so give rise to what are

known as cases of protective mimicry. Again, when such variations arise in the course of experiments on the rearing of larvæ, many would regard them as directly produced by variations in temperature. The fact that a British species may be yellow, orange, or red suggests that the pigment produced depends upon the chemical condition of the cells, and that this may be only indirectly influenced by the state of the temperature. There can at least be no doubt that such variations are of much importance in considering the bearing of the colours of Lepidoptera upon the general problems of evolution.

Mr. Bateson also mentions interesting cases of variations from red to blue or from blue to red in insects, *e.g.* in the wings of butterflies and the tibiæ of locusts. This is especially interesting because blue in insects seems to be always (?) an optical colour, so that the variation in this case must be associated with the development or suppression of surface sculpturing. It is somewhat curious to note that in the same paragraph Mr. Bateson gives as other examples of alternations between blue and red that of various Copepods and of flowers, *e.g.* pimpernel, in both these cases the colours are produced by pigments, which exist in red and blue forms (see pp. 68, 126).

2. *Artificially produced.*—Another question which has attracted much interest in insects is the relation existing between the normal colours of insects showing much variation of tint, and the colours which may be produced by subjecting the developing organism to varying environmental conditions. The subject has of course been especially discussed in the case of the Lepidoptera, and although the whole

question of artificially produced colour-change is too
vast to be entered into in detail here, this chapter
would be incomplete without some reference to it.
As the observations of Mr. Poulton and others have
rendered the adaptive changes of caterpillars tolerably
familiar in this country we shall not discuss them,
but rather note some points in connection with the
changes produced in the butterflies of the genus
Vanessa by subjecting the pupæ to varying conditions of temperature ; these being not quite so well
known.

The occurrence of seasonal dimorphism in the
butterfly formerly described under the two specific
names of *Vanessa prorsa* and *V. levana* is well known,
and the fact that similar colour-variation may be
produced in other Vanessæ by subjecting the pupa
to varying temperatures is also familiar through the
researches of Weismann, Fischer, and others. The
interpretation of the results obtained has always,
however, been a matter of considerable difficulty.
Weismann's conclusion in the case of *Vanessa levana-
prorsa* is, or was, that each pupa contained within it
the potentialities of both forms, and that the temperature was not the efficient cause of the one produced,
but merely the stimulus which set in motion the
necessary pre-existing mechanism. Fischer made
an extended series of observations on the same
subject, and in the main agrees with Weismann,
while Urech (1896) dissents very strongly from this
position. We cannot here enter into detail on the
subject of Urech's paper—it is largely theoretical in
nature—but may mention a few of his points. His
main object is to prove that the differences between

the forms are not due to the action of unknown forces—Weismann's ids—but are dependent upon simple chemical and physical causes, which have a direct action upon the organism.

In the first place, Urech draws a sharp distinction between pigmental and optical colours in the Vanessæ; according to him the bright blue or violet colours only occur in scales which are almost devoid of pigment, so that an actual decrease in amount of pigment may produce an increase in colour-brilliancy. This of course assumes that any scale which is devoid of pigment may display optical colours, and this remains to be proved. Urech also believes that artificially-produced colour-variations are not accompanied by variation in the amount of pigment present in the wings, but only by variations in the distribution of the pigment, resulting in alterations in the extent of the blue patches. For details of this position reference must be made to the original paper.

Urech further makes an interesting suggestion as to the nature of the association between red and yellow pigments. He believes that the pigments are complex organic compounds which owe their colour to the presence of a certain radicle; successive substitutions might then produce pigments of deepening colour. The production of red and yellow pigments from an orange one might, he suggests, occur in the following way :—Let $M{<}^R_h$ be an orange pigment, R being the colour-producing radicle, then

$$2M{<}^R_h = M{<}^R_R + M{<}^R_h \; ;$$

N

that is, two molecules of the orange pigment might combine, and form one molecule of red and one of yellow pigment. The chief interest of this suggestion lies in the fact that such a splitting of colours has been described in the males as compared with the females in both birds and butterflies. It may quite well be that this is in reality due in some such way to a chemical change. In butterflies, as we have already seen, there is certainly a strong tendency for colours like yellow, orange, and red to show oscillation in amount.

Into the numerous other problems connected with the colours of insects space does not permit us to enter here. Many of them are indeed subjects which can only be adequately treated by ento-mologists, and reference should be made to the works of Wallace, Poulton, Weismann, Meldola, Beddard, and others for details. It is, however, hoped that the foregoing summary will be of use in calling attention to aspects of the subject which are less familiar, but are undoubtedly of great importance.

THE COLOURS OF SPIDERS

Among other Arthropods the colours of spiders merit at least a brief notice. In them the pigments do not seem to have been investigated at all, and the structural colours only to a very slight degree, but there are some interesting observations on the colours themselves.

As compared with insects, the characteristics of spiders which are important for our purpose are the practical absence of obvious segmentation in the abdomen, and the delicacy of the cuticle which covers that region. The whole surface of the body is not infrequently covered with a dense coating of hairs, which are analogous to the scales of Lepidoptera, and are often important in coloration. According to Dr. Henry M'Cook (*American Spiders*), the cuticle only contains a relatively small amount of pigment, this being chiefly confined to the "soft skin " (= epidermis ?).

(1) *Optical Colours.*—Opinions differ greatly as to the prevalence of bright colours in the group, non-specialists being generally inclined to regard spiders as dull-coloured, while enthusiasts like M'Cook and the Peckhams speak of them as rivalling insects and birds. It is at least certain that metallic colours do occur in various species. The commonest of these colours appears to be a metallic or silvery white due to the structure of the hairs, but in some cases these hairs or scales give rise to brilliant iridescence, or to pure blue and green colours. In certain instances, as for example in the mandibles of *Phidippus morsitans*, the cuticle itself is, as in many beetles, finely ridged, and so gives rise to a brilliant green colour. In this connection it is interesting to note that the Peckhams remark that the bright colours of the male are most conspicuous in the anterior region. This they ascribe to sexual selection, but if the colours are structural they can probably only occur in those regions where the cuticle is markedly differentiated, and these regions are especially the appendages of the mouth and the neighbouring parts.

(2) *Pigmental Colours.*—Apart from structural colours, those due to pigment seem to be bright and

varied ; black, red, green, yellow, brown, etc., are apparently all common. Although there is not the sharp division of the life-history into two contrasted stages which is so obvious in many insects, yet there is in many cases a marked development of colour in ontogeny. The following facts are taken from Dr. M'Cook's book quoted above.

Development of Colour.—Almost all very young spiders are light yellow or greenish-white ; as development proceeds the colours deepen and yellows and browns appear, but it is not until the spiders begin to weave webs on their own account that the colours characteristic of the species develop. The characteristic patterns may be present in the young, but differ in colour from those of the adult ; thus *Argiope cophinaria* has pure white markings in the situations in which the adult has yellow markings. A general deepening of tint is markedly characteristic of the passage from youth to maturity ; and this may take place to such an extent that the markings which are at first distinct may gradually disappear. As illustrations of the deepening of colour we may take *Epeira trifolium*, which is at first white, then becomes yellow, at full maturity displays brilliant and very variable colouring, and then after laying her eggs becomes a dull, dark colour, a change which immediately precedes death. A somewhat similar series of colour-changes is displayed by *Tegenaria medicinalis*, which is first pale, then deep yellow, and finally blackish. In marked contrast to these cases, however, there are a few forms, like *Epeira strix*, which are deep black at the time of hatching. The last-named species is very variable in colour, but the adults are usually yellowish with red bands, or sometimes whitish ; the colour-development is thus in marked contrast to the cases already described.

Before passing on to discuss the beautiful and complex markings seen in many spiders, we may dwell for a little on some cases of colour-variation more complex than those merely due to development, and on the colours of the sexes.

Variation in Colour.—We have already described the colour-changes occurring during development in *Epeira trifolium*, but the adults themselves show an extraordinary variability, often changing colour markedly in captivity. The specific name is founded upon the resemblance of certain markings on the dorsal surface of the abdomen to the leaves of trefoil, but it is only in certain cases that these markings are at all distinct. The following list is compiled from Dr. M'Cook to illustrate the common colour-variations :—

Colours of Body.	Colours of Legs.
1. White with faint black markings.	Brown and white.
2. Do. do.	Black and white.
3. Orange to crimson - red with yellow markings.	Dark brown and white.
4. Bright red with yellow markings.	Yellow and red-brown.
5. Yellow to brown with white markings.	Orange and brown.
6. Yellow to orange.	Yellow and brown.

These interesting variations, which show such a close connection between white, yellow, red, and brown, suggest strongly that there is a relation between the pigments corresponding to that which exists in the

Lepidoptera between white, yellow, and red pigments, but exact observations have still to be made.

Sexual Coloration.—As to the colours of the sexes, the careful observations of Mr. and Mrs. Peckham have familiarised us with the fact that the males are not infrequently more brightly coloured than the females, the bright colours being especially marked on those regions of the body which are most fully displayed during the antics of courtship. The males are not, however, invariably brighter than the females. Thus in *Nephila plumipes* the male is dull brown and small, the female has a jet-black cephalothoracic region largely covered with silvery hairs, while the abdomen is an olive-brown colour tending to light yellow above and with yellow and white spots and stripes. The legs are yellow with dull red rings, several of the joints being furnished with plumose tufts. This is an interesting case, because if the colours and ornamental tufts had been present in the male, they would have been certainly ascribed to sexual selection. As it is, male spiders are asserted on all sides to prefer amiability to beauty, so that such an explanation is hardly valid.

Markings of Spiders.—The last characteristic of the coloration of spiders which we must consider is the beauty and complexity of the markings. These markings tend especially to occur on the abdomen, where they are usually very complex ; the cephalothorax is usually more uniform, but the legs not infrequently display bands of colour of a deeper tint than that of their ground-colour. The abdominal markings often take a leaf-like shape, as, for example, is well seen in the " folia " which give its specific

name to *Epeira trifolium*. The relative uniformity
of the cephalothorax is perhaps in part associated
with·the uniform thickness of its cuticle.   The markings of the abdomen are, as we shall see later in more
detail, probably associated with the tendency of the
epidermal pigment to become aggregated round the
little pits which indicate the points of attachment of
the muscles.   M'Cook considers that the variation
in colour and markings which is so marked in the
abdomens of many spiders, may be due in part to
variations in the amount of distention in this region
due to the presence of eggs and so on.

   Dr. M'Cook's volumes contain many other interesting observations on the relations of colour to habitat
in spiders, on instances of so-called mimicry, etc., for
which reference must be made to the original.   There
are various observations in the literature of the subject
tending to prove that spiders in some instances can
vary their tints according to their environment ; that
harmony with the colours of the environment which
is known as " protective coloration " seems also to be
relatively common.

CHAPTER IX

THE COLORATION OF MOLLUSCA AND OF INVERTEBRATES IN GENERAL ·

Colours of the Mollusca : (1) of the Mantle, (2) of the Shell, (3) of the Secretions—Pigments of the Mollusca—Green Oysters—Characters of Coloration of Invertebrates—Patterns and Markings of Organisms ; Plants, Cœlentera, Leeches, Arthropoda, Mollusca, Vertebrates—Meaning of Patterns.

THE Mollusca are especially remarkable for the number and brilliancy of their pigments ; unfortunately, however, these have been relatively little, investigated. The pigments occur especially under three circumstances—in the mantle, in the shell, and in the secretions poured out by many, of which the far-famed fluid of *Purpura* may be taken as a type.

1. *Colours of the Mantle.*—The mantle-folds themselves, or the general body wall, are not infrequently highly coloured, and this whether the coloured regions are visible during life or not. Thus Mr. Saville Kent describes the mantle of *Tridacna compressa*, the Frilled Clam, as being during life resplen-

dent with the most gorgeous colours. This giant clam occurs freely among the corals of the Australian Barrier Reef, and during life the valves gape widely, disclosing the frilled mantle-folds which give the species its popular name. These are often coloured with shades of blue varying from " palest turquoise to the richest ultramarine," or are green variegated with spots and markings of black. Sometimes the ground colour is purple or rich brown spotted and streaked with bright blue or green. The shell in this clam is a pale yellow, and the organisms are exceedingly conspicuous on the reefs. In many Molluscs the pigments of the mantle are confined to special spots, as in the case of the simple " eyes " of many Lamellibranchs, but cases of uniformly diffused colour are far from uncommon. The delicate pink colouring of the fringed mantle-folds of *Lima* is familiar to all those who have done coast dredging. In *Patella vulgata*, the common limpet, the dorsal body wall is of a deep blue-black colour, while in the little tortoise-shell limpet (*Acmæa testudinalis*), common on Scotch coasts, the same region is a delicate blue-green. In the shell-less Gasteropods, such as the sea-slugs (*Doris*, etc.) and the terrestrial slugs *Limax*, etc., the whole of the body wall is usually coloured, often with bright pigments, but this is much less inexplicable than the occurrence in the mantle of shelled forms of bright pigments having no obvious relation to the pigments of the shell.

As to the colours of the Cephalopoda there is little need either for description or eulogy ; most people must have seen and admired the lovely

changing tints of the hapless cuttles which the storms
of spring cast upon the beach. To those who have
not, no learned talk of chromatophores will suggest
the delicate rosy flush which comes and fades as one
touches the sensitive skin.

The cuttles of our own shores are chiefly remarkable for their changing tints rather than for their
brilliancy, but some tropical forms display very
bright colours. Thus *Octopus pictus* is of a yellowish
colour with spots " suffused with shades of the
richest ultramarine blue, each spot having a light
centre and a dark annular border " (Saville Kent).
According to Agassiz the deep-sea cuttles are usually
brown or purplish-brown.

2. *Colours of Shells.*—The fact that most shells
are distinguished by their rich and beautiful colouring is familiar enough ; their variety and beauty of
marking has endeared them alike to the savage, the
child, and the artist. Brightness of colour is most
common among the shallow-water forms, especially
of the tropics ; deep-sea forms tend to be pale in
colour, but often possess an iridescent sheen absent
from those living in shallow water (Agassiz). Many
shells are coloured on the inner surface as well as the
outer, and the colouring-matter is rarely uniformly
diffused, being most frequently arranged to form
patterns and markings. According to Dr. Woodward,
" those which are habitually fixed or stationary (like
Spondylus and *Pecten pleuronectes*) have the upper
valve richly tinted whilst the lower one is colourless."
In some cases the mantle is coloured and marked in
the same way as the shell. In *Ianthina* the upturned
base of the shell is coloured by the violet pigment

which is present in the well-known secretion of the
animal.   Among the bright colours of shells shades
of red, orange, and yellow are perhaps the most
common, but blue and green also occur.

3. *Coloured Secretions.*—The third way in which
conspicuous colours occur in the Mollusca is in connection with the secretions poured out by many.
Among these may be mentioned the ink of Cephalopods, the violet fluids of *Ianthina* and *Aplysia*, the
colourless fluids of species of *Murex* and *Purpura*
which in air undergo a series of changes rendering
them ultimately violet-coloured, the purple fluid of
species of *Scalaria*, and so on.   Many of these colouring-matters are, or were, of commercial value, and it
is somewhat interesting to note that in animals, as in
plants, it is usually the pigments which are inconspicuous during life which can be utilised as dyes.

PIGMENTS OF THE MOLLUSCA

To pass now to the pigments themselves : we find
that these are exceedingly numerous and varied ; in
this respect they are sharply contrasted with the
Crustacea, whose pigments are very uniform throughout large groups.   Many of the pigments are, however,
little known.   The Mollusca agree with the Crustacea
in their blood pigments, and in the presence of a
limy exoskeleton, frequently impregnated with lipochrome pigments.   In both groups the blood may
contain hæmocyanin or hæmoglobin.   The latter, rare
in the Crustacea, is distributed in an exceedingly
capricious way throughout the Mollusca, sometimes
occurring in the hæmolymph, sometimes only in

certain muscles; but in the Mollusca, as else-where in Invertebrates, it demands further study. Hæmocyanin, on the other hand, is very widely distributed in both groups; the question whether it can function as a pigment strictly so called is a somewhat difficult one. It will be recollected that it differs from a pigment like hæmoglobin in dis-playing colour only in the oxidised condition—reduced hæmocyanin is quite colourless. It is in consequence difficult to believe that hæmocyanin can be of permanent value as a pigment, except in delicate organs freely exposed to sea-water. Under such conditions, however, there is nothing intrinsically improbable in the idea that it may give rise to brilliant colour. In dissecting recently killed sea-slugs the bright blue colour of the abundant blood is very noticeable. It seems quite possible that the bright blue tints seen in the transparent papillæ of many Eolids may be due to the oxidised blood shining through. The colours are at least exceedingly fugitive after death.

The fate of hæmocyanin in the body is unknown, but there is no reason to suppose that it can give rise to series of pigments in any way resembling those which arise in vertebrates from the breaking down of hæmoglobin. Derivatives of hæmoglobin have on the other hand been described in various mol-luscs. Dr. M'Munn, for example, describes hæmato-porphyrin in the slugs *Limax* and *Arion*, and so on. According to Krukenberg the shells of some species of *Haliotis*, *Trochus*, and *Turbo* are coloured by biliverdin, one of the bile pigments of Vertebrates, while other species of the first and last genera have their shells coloured by turbobrunin, a red pigment easily converted into biliverdin.   Krukenberg regards this as evidence that the bile pigments can arise independently in Molluscs, without the intervention of hæmoglobin, but the whole subject is still obscure.

As colouring agents of shells the ubiquitous lipochromes also occur; they have been found in such shells as the gaily coloured *Pectens* and in *Littorina ;* it is probable that here, as in Crustacea, they form compounds with the lime of the shells, and are most stable in those cases where the shell contains the largest amount of lime.   Numerous other pigments have been described as colouring the shells of Molluscs, but as most of them are very imperfectly known, it is unnecessary here to mention the names which have been given them.   The relation between the pigments of the mantle and the shell, a subject of great interest, does not appear to have been investigated at all.   Although in some cases the two structures are similarly coloured, yet in others there is no apparent relation.   It is quite probable that this may often be due to the fact that the pigments of the shell are compounds or oxidised products, but as yet nothing seems known on the subject.

In connection with this subject we may mention the peculiarly vivid pigment of the tortoise-shell limpet.   This pigment does not seem to have been described, in spite of its striking tint.   It is entirely confined to the cells of the ciliated epithelium which lines the shell, is turned in to cover the dorsal surface of the foot, and also covers the mantle-fold.   During life the colour is very inconspicuous, being only visible in the mantle-fold and in small specimens as a faint green line down the centre of the foot, produced by the coloured dorsal epithelium shining through. In the cells the pigment is present in small granules, varying in colour from light green to greenish-blue or even light blue. At times it has a black or brownish tint, the extreme edge of the mantle-fold being always brown. The pigment is soluble in water, or better in alcohol, but in solution the colour rapidly disappears. Alkalies destroy the colour at once, but it returns upon acidification ; dilute acids give the pigment a blue tint. The shell of this little limpet is a warm brown colour, and in view of the tendency of the green pigment to become brown, it is not unreasonable to suppose that during the process of shell-formation the green pigment becomes converted into the brown.

This green pigment is of some interest ; it varies, as already seen, from blue to green, and is perhaps somewhat widely distributed among Invertebrates. The one just described certainly resembles closely that found in the eggs of *Eulalia viridis*, and in *Thalassema*. The origin of the pigment is quite unknown, but it is not impossible that it arises from the "enterochlorophyll" of the digestive gland.

As to the glandular secretions which are so common in the Mollusca, we find that their pigments exhibit very diverse characters. The ink of Cephalopods, for example, contains a black or brown pigment, which is nitrogenous and is said to have a chemical composition almost identical with that of the black pigment of crows' feathers. According to the older analyses of Girod it is free from ash, but Dr. Emil André considers it identical with melaine, the pigment of *Limnæa stagnalis*, which he states is iron-containing.   André regards melaine as a waste product stored up in the tissues.   When we consider the almost universal presence of pigment in the shell of shelled forms, it seems not unnatural to conclude that the large amount of pigment in the ink of Cephalopods is associated with the suppression of the shell, which has thus, as it were, forced the organism to get rid of its pigment by some new means.

The beautiful purple colouring-matter of the fluid of *Ianthina* has always attracted much attention, but in this case, as in those of the chromogen of *Murex* and the pigment of *Aplysia*, there have been few exact investigations.   The colour of the secretion of *Aplysia* has been asserted to be due to natural aniline dyes, but this, according to Krukenberg, is incorrect.   Moseley found it to be soluble in alcohol with a purple colour, turning violet with acid ; it was very unstable and gave two sets of spectra according as it was or was not acidified.   The pigment of *Ianthina*, on the other hand, dissolved in alcohol to form a violet or pinkish-blue solution, acid turning it a light pale blue.   It gave a spectrum with three bands.   It would be interesting to know if these curious pigments occur elsewhere in Mollusca, and if they have any connection with the blue pigments of shells like the common mussel.

In connection with the pigments of the secretions we may note the peculiar pigment enterochlorophyll, which is so widely spread in the digestive glands of the Mollusca.   In *Patella* I find that it occurs in the epithelial cells lining the alimentary canal, in the digestive gland and its secretion, intermingled with the contents of the gut, and finally in unaltered form in the fæces. It thus acts like a true bile pigment. We have already frequently spoken of the peculiar group of pigments to which enterochlorophyll belongs, and of the facility with which they can be made to yield bright-coloured derivatives. It is evidently here a useless substance, and it is very probable that in some cases, instead of being eliminated unaltered in the fæces, it may undergo modification into other pigments which may be deposited in the mantle or shell. We have already noticed such a suggestion in the case of *Acmæa*, and it is possible that many of the bright pigments of Mollusca arise in this way.

The yellow, orange, or black pigments of naked forms such as the slugs, *Doris*, etc., are probably due either to lipochromes or the dark-coloured " melaines."

It seems probable that the pigments of Cephalopoda are chiefly of the dark-coloured nitrogenous type, though they have not been fully investigated. The beautiful changing colours are in large part due to the movements of the chromatophores, which as they expand or contract alter the whole appearance of the animal. So long ago as 1852 Brücke watched the colour-changes of *Sepiola rondeletii*, and noting how the tints varied in the order of the spectrum from blue to green, yellow, and red, came to the conclusion that the colours were optical. He thought that they were due to the colours of thin plates, but according to Krukenberg they are due to fine ridges in the surface of the cells.

In connection with the colours of Mollusca, we may mention the vexed question of the cause of the colour of Green Oysters. Prof. E. R. Lankester, who studied the question in 1886, described the green pigment as occurring especially in amœboid cells, which he speaks of as "crawling over the surface of the gills." He found the pigment to be very refractory to solvents, to be of a blue-green colour, and not to contain iron or copper. As a pigment of similar nature occurs in *Navicula ostrearia*, a diatom found in the tanks in which Marennes oysters occur, and used by them as food, he regarded the pigment (marennin) of the green oysters as a derived pigment, which was removed from the gut by amœbocytes and ultimately transported to the gills and there eliminated. MM. Chatin and Muntz in 1894 investigated the distribution of iron in the oysters, and found that the green parts contained about twice as much iron as the colourless parts, and that the amount of iron varied with the intensity of the pigmentation. More recently Dr. Carazzi, in reinvestigating the whole subject, has come to the following conclusions. The green pigment occurs in the branchial epithelium and in amœbocytes included in this epithelium, but not in the "gland-cells" of Lankester. It is not transported by amœbocytes from the gut, is not due to *Navicula*, but is a special product of metabolism in the oyster, perhaps nutritive, and is a peculiar organic compound containing iron. If these con-

O

clusions be correct, then marennin must be added to the list of the pigments of Mollusca. In this connection it is interesting to note its extreme stability, a character which is rare among green pigments in Mollusca or elsewhere ; the extreme vividness of the tint is also very striking.

Carazzi's statements, if well established, will overthrow the view that the green oysters are coloured by derived pigment, which is a point of some importance. Of course there are other instances, *e.g.* that of Mr. Poulton's caterpillars, which seem to be well established and to prove the possibility of pigment transference from one organism to another, but we have still no means of knowing whether or not this is of common occurrence in the animal kingdom. The question is interesting, because suggestions as to pigment transference have been very freely made to explain cases of resemblance between organisms and their surroundings.

The subject of "Green Oysters" has given rise to so many heated discussions that, although a little apart from our main theme, we may pause to note a very interesting paper by Drs. Boyce and Herdman on an abnormal green pigmentation. Part of the interest taken in the general subject has been due to the fact that there have repeatedly occurred instances of poisoning from the eating of green oysters, where the symptoms have been declared to be those of copper - poisoning. On the other hand, many examinations of coloured oysters have failed to demonstrate the presence of copper. The researches of the authors mentioned have shown that the dispute presents a strong analogy to that historic

one waged over the question whether lobsters are
black or red !   They find that, altogether apart from
the famous Marennes oysters, there occurs in diseased
oysters a green colour due to the presence of a large
amount of copper.   Drs. Boyce and Herdman do
not believe that the large increase in the amount
of copper normally present is necessarily the result
of the presence of copper in the food, but " are
inclined to suggest that it may be due to a disturbed
metabolism, whereby the normal copper of the hæmocyanin, which is probably passing through the body
in minute amounts, ceases to be removed, and so
becomes stored up in certain cells."   There is every
reason to believe that such " green oysters " are
highly poisonous, while the Marennes oysters are of
course quite harmless.   The paper thus settles a
time-honoured controversy in the most satisfactory
manner.

CHARACTERS OF THE COLORATION OF INVERTEBRATES

With the Mollusca we end our study of the colours
and colouring-matters of the Invertebrates.   As a
whole the Invertebrates are characterised by the
number and diversity of their pigments, which in the
simple forms occur in connection with internal organs
and are often very variable ; by the frequent beauty of
their structural colours ; and in the case of the more
differentiated forms in certain groups, by the beauty
of their patterns and markings.   In Vertebrates the
pigments are less numerous, often less vivid, and the
beauty is usually due either to structural colour or

to the markings.   It is fitting, therefore, that at this
point, before passing from the Invertebrates to the
Vertebrates, we should consider for a little the meaning and characters of patterns.   We have again and
again encountered such types of coloration among
the Invertebrates, but when we pass to Vertebrates
we shall find that absolute uniformity of colour is
relatively rare, and that in most cases the markings
or patterns are the conspicuous features.   We propose, therefore, to consider here first some of the
characters of the colour-patterns naturally occurring
in organisms, and then some suggestions as to their
meaning.

THE PATTERNS AND MARKINGS OF ORGANISMS

Plants.—Among plants colour-markings occur
most conspicuously in flowers, for the artistic value
of foliage depends in the general case upon simple
colour contrasts between stem and leaves and upon
beauty of form rather than upon complexity of
marking.   In flowers, as is obvious upon a moment's
reflection, the markings depend greatly upon the
form and character of the parts.   Flowers in which
the floral envelopes consist of similar parts have these
parts similarly marked, as, for instance, the stripes or
markings on the petals of tulips, of geraniums, snow-
drops, and so on.   Slight irregularities of colour are
associated with correspondingly slight variations in
structure, as seen in some of the so-called honey-
guides, such as the spot of dark colour in the flower
of the rhododendron.   On the other hand, the more
complicated colour-markings are usually confined to

one region of the corolla, and are associated with irregularity of parts. This is well seen in the markings of orchids, of the foxglove, of pentstemon, and many others, in which cases the parts of the corolla are irregular, and the markings are confined to certain of them. In general, in flowers the markings tend to be similar in homologous parts, but variations in the structure of homologous parts tend to be accompanied by variation in the colour-patterns.

Cœlentera.—The colours of Protozoa and sponges are too simple to show any arrangement into patterns, or perhaps it would be more accurate to say that the structure is too simple. In any case, we do not find the phenomenon markedly displayed before we reach the Cœlentera. In them it exhibits itself with the same simplicity seen in plants. The organisms have the radial symmetry so characteristic of flowers, and again we find· that homologous parts show similar colouring. One tentacle is in structure and development like another, and the colours, whether uniform or arranged in simple banded patterns, tend to be identical throughout. As we have already seen, structural differentiation and the development of new colours tend to be associated; thus knobbed or capitate tentacles usually have the terminal knob of a different colour from the rest of the tentacle—a fact which may. be susceptible of a simple physical explanation. Again, when the tentacles are greatly reduced in size, they may cease to behave as individuals in the general colour-scheme, which has then as its unit groups of tentacles. This occurs in the giant anemones (*Discosoma*), and is of interest . because the subordination of the individual coloured structures to the pattern of the whole is so common among higher organisms.

Leeches.—Among worms, as we have already seen, the leeches are distinguished *par excellence* by their markings and patterns. There are few facts more striking to one interested in colour than the change seen in passing from the Nemerteans and the marine Annelids, as one finds them on the shore, with their bright, uniform, and fugitive colours, to the leeches, with their dark persistent tints and their beautiful markings. That the colours of, for example, the medicinal leech show a considerable amount of variation, no one accustomed to handling specimens can deny, but at the same time there is sufficient constancy to admit of the dark spots being employed as a ready means of counting the segments. The occurrence of a constant and elaborate scheme of colour-markings at such a low grade in evolution is very interesting.

We may notice also that the development of elaborate patterns in the leeches, as compared with the marine worms, is associated with the development of a large amount of dark pigments. This is interesting because it is perhaps universally true that elaborate patterns are dependent, at least in part, upon dark pigments, while the bright pigments tend as a rule to be more uniformly distributed. It is difficult to avoid coming to the conclusion that the fact is associated with the insolubility of the dark pigments, which will render them on the whole less readily diffused than the more soluble bright-coloured pigments. Lipochromes, for example, dissolve readily in solutions containing albumen or in fat, which seems to account in some degree for their usual uniform distribution in the tissues or structures in which they occur. We may note here that in the only case in which a dark (melanin) pigment has been subjected to a thorough examination, it was found that the insolubility was not intrinsic, but was due to intermixed substances which acted as mordants; if, however, such substances tend always to occur in association with the pigment, the practical result is the same. The question whether the numerous dark pigments found in the animal kingdom are related to one another is an interesting one, but has not yet been properly attacked.

It should be noticed in leeches that in general terms each segment tends to repeat the colour, as it does the structure, of the others. The segments are here the similar parts which tend to resemble one another; we have already (p. 108) considered the ingenious theory which seeks to correlate the colour with the detailed structure of the muscles.

Arthropoda.—In the Arthropoda we have again the same contrast between uniformly, often brightly, coloured organisms on the one hand, and on the other those showing wonderful arrangements of colour-markings, with black or brown as a basis. In a rough way the contrast again lies between uniformly coloured marine forms—the Crustacea, and patterned terrestrial forms—the Insects. The contrast is only roughly true, however, for we have already seen how many insects there are, especially in the less specialised orders, which are uniformly coloured in greens or browns.

As to the patterns and markings of insects, if we

take caterpillars, dragon-flies, bees, or other cases where the segmentation of the body is obvious, we find that, just as in leeches, the coloration bears a close relation to the segments, as is seen in the banding of many caterpillars, of the abdomen of dragon-flies, and so on. But since the Arthropods, as compared with worms, show a subordination of the individuality of the segments to the needs of the whole—a synthesis of segments—so we find that there is a constant tendency for the pattern to dominate the whole organism instead of being the result of the patterns of the segments. The whole is symmetrical round a median line, but is in most cases not merely due to the repetition of the patterns of the segments. This is especially well seen in caterpillars in the colouring of, for example, the head. This is a specialised region of the body often with much thickened cuticle, and accordingly we find that it often differs in coloration from the rest of the body. When we pass from caterpillars to forms like dragon-flies or bees, we find that, while the relation of the pattern to the segments is obvious in the relatively un-specialised abdomen, it is lost in the much modified thorax. In other bees again, as in some of the humble bees, the coloration has largely lost, even in the abdominal region, any direct relation to the segmentation. In this connection the coloration of spiders offers some points of great interest. As we have already seen, the markings, especially on the abdominal region, are often exceedingly complex. According to M'Cook, the markings tend to adopt a leaf-like shape, and these *folia* appear to be related to the little pits on the surface of the abdomen which mark the internal attachment of the muscles. These spots are " centres for aggregation of pigment," and have been supposed to indicate the segmentation of the body, and thus, as Dr. M'Cook says, the patterns probably bear a direct relation to the segmentation. The same facts are observable to a much less degree in the cephalothorax, and M'Cook suggests that the annuli round the legs, which occur especially in the neighbourhood of the joints, may be determined in a similar way by the arrangement of the muscles. All this is exceedingly interesting and suggestive, but from a study of other forms, *e.g.* insects, one would expect to find that the gradual disappearance of marked segmentation in spiders was associated with a profound modification of the coloration. If, as there is reason to believe, coloration and structure are directly associated, then the new structural characters of the body should be reflected in the markings. It seems probable that this really occurs. Thus *Argiope argyraspis* has on the abdomen two longitudinal yellow lines and thirteen unbroken transverse black lines as well as other incomplete ones. Now as the embryos in spiders are commonly supposed to have ten to twelve abdominal segments at most, and as the posterior ones degenerate during development, it is at least unlikely that these transverse bars have a direct relation to segmentation. It seems more probable that we have another instance of the fact that in specialised organisms the process of integration which has so profoundly modified the segmentation has been accompanied by changes in the colour-markings.

In butterflies, beetles, and other insects with large

and conspicuous wings, and small or concealed abdomen, the colour-markings occur on the wings or wing-covers, and have thus no relation to segmentation. In this case the coloured structures are lateral organs and not parts arranged in linear series, but they conform again to the rule that organs occupying similar positions display similar coloration ; the two fore-wings display similar colours, and the hind agree with one another and may agree with the fore-wings. On these points there are some interesting observations · by Mr. Scudder, whose statements have been confirmed and amplified by Mr. A. G. Mayer. In discussing the butterflies of North America, Scudder showed that the number of instances in which similar markings appear in the same areas in the two pairs of wings is too large to be due merely to coincidences. The process is most readily traced in the case of ocelli, which usually tend to be similar in size and position, and to be situated between the same branches of homologous veins. In summing up his own and Scudder's observations, Mayer (1897) lays down the following statements with regard to colour-patterns in butterflies :—

1. "Any spot found upon the wings tends to be bilaterally symmetrical both as regards form and colour, the axis of symmetry being a line passing through the centre of the interspace in which the spot is found, and parallel to the direction of the longitudinal nervures."

2. "Spots tend to appear not in one interspace only, but as a row occupying homologous places in successive interspaces."

It is thus seen that the wings and their homologous parts tend to behave in the colour scheme as do the successive segments in little differentiated forms. So close, however, is the relation between structure and colour, that differential growth of the wings resulting in the overlapping of parts is also accompanied by corresponding change of pattern. This is well seen in many butterflies where the fore-wings overlap the hind. We have already noticed the peculiar case described by Sharp in which the fore-wings of an orthopterous insect differed from one another in their colouring, but the irregularity was corrected in the position of rest by the overlapping of parts. It is a curious example of that detailed relation which exists between colour-markings and structure.

In spite of the conspicuousness of colour-markings in many insects, there seems as yet no certainty as to the proximate cause or meaning of the patterns. Over the markings of caterpillars, their meaning and their evolution, it is true that many fierce controversies have raged ; but the questions as to whether longitudinal or transverse markings are the most primitive, the number of the possible longitudinal stripes, their use, and the kindred questions seem to be decided so much by the caprice of the individual investigator, that a non-specialist may be permitted to retire from the field until the parties concerned have been able to patch up some sort of a truce.

Mollusca.—In the Mollusca colour-patterns are, except in the cuttles, confined to the shells, in which banding is very common, and is well seen in common snails (*Helix*). More elaborate schemes of ornamentation are not uncommon, and are not infrequently associated with a sculpturing of the shell. According to the Countess Maria von Linden, both the sculpturing and the colours show a definite orderly progression in which phylogeny and ontogeny run parallel. In the garden snails, *Helix nemoralis* and *H. hortensis*, there is an extraordinary amount of variation in the bands, but nothing is known of the reason.

Vertebrates.—So far we have seen that colour-patterns are best developed in animals in which there is a distinct segmentation of the body, that in unspecialised forms they show a direct relation to the segments, but that as specialisation proceeds, they tend to lose this simple and direct relation. When we pass upwards to Vertebrates, we find that an apparent relation to segmentation is completely lost. It is true that in the case of the ringed snake, as we shall see, there appears to be a direct relation between the ornamentation and the arrangement of the cutaneous blood-vessels, and these may well have a segmental significance, but at the same time the statement on the whole is true that a direct relation between the markings and the segmentation of the Vertebrate body is no longer obvious. In the general case the pattern in a Vertebrate dominates the whole organism and is not produced by the repetition of one colour scheme. This is perhaps an advance due to the perfect synthesis of the body, just as the frequent striking difference between the fore and hind wings of a butterfly is an advance when compared with the segmental patterns of caterpillars ; or it may be an indication that the segmentation of a Vertebrate is in essence a different thing from that of a worm or an insect.

Though there is no successional repetition in a Vertebrate, the rule that parts occupying similar positions tend to be similarly coloured holds good here as in Invertebrates. The similarity seems, however, here to have relation to position with reference to the median axis of the body only. A brief study of birds' feathers affords very convincing proof of this. Thus, though the secondary quills of one wing in a bird showing complicated feather-markings resemble those of the other wing, it will be found that the quills are not all absolutely the same, but form a graduated series, of which the first and last members may differ very considerably. The usual difference between the markings of the two sides of the vane may be explained in wing-quills as due to the unequal development of the two sides, but, except in the case of the central rectrices, it is quite as common in the tail-quills, in which the two sides are equally developed. In short, it is a general fact that in Vertebrates, where the coloration depends upon the colours of the epidermal outgrowths, the pattern has reference to the body as a whole, the colours of the individual structures being completely subordinated to that whole.

As to the nature and meaning of patterns, there can be little doubt that in simple organisms they are closely related to segmentation, are expressions of the same process. It has been dimly suggested by various authors, but nowhere with such force and eloquence as by Mr. Bateson, that both patterns and segmentation may be purely mechanical phenomena, and that " the perfection and symmetry of the process, whether in type or in variety, may be an expression of the fact that the forms of the type or the variety represent positions in which the forces of division are in a condition of mechanical stability" (*Variation*, p. 71). It can hardly be doubted that this explanation has an important bearing on those simple forms of patterns which are directly related to segmentation, but even the approximate explanation of the complicated colour-markings of Vertebrates seems as yet impossible. "Forces of division" seems too vague an expression to be of much aid in their case. For some other explanations of markings in Vertebrates, reference should be made to Mr. Wallace's *Darwinism*.

CHAPTER X

THE COLOURS OF FISHES

The Colours of the Lower Vertebrates—General Characteristics of Vertebrate Coloration—Colours of Fishes, their Characters and Distribution in Tropical, Abyssal, and Temperate Forms—Coloration of Flat-fishes—The Pigments and Structural Colours of Fishes.

OF Vertebrates below fishes little need be said as to colour or pigments. The Tunicates alone display any marked colouring, and are often bright and varied in tint. The pelagic forms are mostly transparent and almost colourless, and, as we have seen, they are often brilliantly phosphorescent. Those of sedentary habit, on the other hand, often show a great range of colour. Krukenberg describes the species of the genera *Didemnum* and *Botryllus*, in the neighbourhood of Trieste, as exhibiting the following colours :—blue, violet, yellow, yellowish-green, orange, brown, and black, and observation on the shore confirms this impression of prevalent brightness of tint.

Krukenberg made a few observations on the pigments which are of interest. The red and yellow colours are apparently due as usual to lipochrome pigments. The blue and violet colouring-matters are very unstable, turning brown with most reagents, including water and dilute alkalies. Curiously enough acid restores the blue colour. Certain of the yellow pigments are not lipochromes, but singularly unstable substances of quite different characters. Thus the lymph of *Ascidia fumigata* is coloured by a yellow pigment which is very soluble in alcohol and ether, and slightly soluble in water. On standing in air, however, the solutions rapidly become dark coloured, and ultimately black. This is, according to Krukenberg, due to ferment action ; the interest of it is, however, that the tunic of this species is marked with black, the pigment being probably the same as that which arises from the alteration of the yellow pigment. It will be remembered that in many lepidopterous larvæ the blood contains a green pigment which darkens in a similar way on exposure to the air, and that this is possibly the origin of the dark pigments both of larvæ and adults. These curious pigments—uranidines of Krukenberg —are worthy of further investigation.

M'Munn (1889) confirms Krukenberg's statement that most of the pigments of Ascidians are of lipochrome nature.

GENERAL CHARACTERISTICS OF VERTEBRATE COLORATION

Before passing on to consider separately the divisions of the higher Vertebrates, it may be well to mention some of the peculiarities of the colours as a whole. In the first place, in spite of the varieties of tint and colouring there is a remarkable uniformity of pigments. With a few exceptions, turacin being perhaps the most important, the pigments seem to be all either lipochromes or melanins, the lipochromes predominating in fishes, amphibians, lizards, and birds, the melanins in snakes and mammals (Krukenberg). As the melanins are often regarded as derivatives of hæmoglobin, and the lipochromes of fats, some would say that there are two kinds of pigments in Vertebrates—effete blood pigments and modified fats (see Kükenthal). Secondly, in many cases the general coloration is due either to structural effects or to the patterns and markings rather than simply to the pigments themselves. The development of structural colours is associated with the degree of development of the cuticular outgrowths, and reaches perhaps its greatest height in birds. In mammals bright structural colours are rare, and the frequent beauty of colour is due solely to the unequal distribution of the melanin pigments which gives rise to bands or spots. Eimer has endeavoured to prove that in all Vertebrates longitudinal markings are the most primitive form of coloration, and that spots, transverse stripes, or uniform coloration are secondary derivatives. His work on the *Markings of Animals* shows, at least, that there is an extraordinary constancy in the markings which run through orders, as is, for example, well seen in the Carnivora among mammals.

Sexual dimorphism of colour is another character which is exceedingly common among Vertebrates. This may manifest itself simply in the greater brightness or purity of tint in the male as in some fishes and many birds; or in increased growth and development of the cuticular structures in the male, associated usually with a great development of optical colours and of pigment; or as in many mammals there may be simply a special development of the exoskeleton, not associated with a great increase of pigment, or any special brilliancy of tint. Beauty in mammals is, however, usually due rather to form than to bright colour.

THE COLOURS OF FISHES

1. *Tropical Fishes.*—Among the colours of fishes, those of the forms inhabiting the neighbourhood of the coral reefs of warm seas must be especially mentioned. We have already seen how brilliant are the tints of the coral polypes themselves, and the same brilliancy seems to occur almost universally among the inhabitants of the waters round the reef. In fishes, however, the brilliant colours are probably produced not by pigments but by structure. The colours fade with extraordinary rapidity after death or removal from the water, and according to those who have seen the fish in their natural condition, the coloration is at best quaint and striking rather than beautiful. Generally speaking, the tints seem to be very vivid and hard, with a lack of light and shade; the occurrence of sharply contrasted bands or spots on a bright ground is exceedingly common. Bright green, blue, red, and yellow seem to be among the commonest colours, and there are some striking instances of sexual dimorphism. Thus in *Ostracion ornata* the male has a ground colour of grass-green,

with spots and stripes of brilliant blue, while the female, often classed as a separate species under the name of *O. aurita*, is of a pale yellow or flesh-colour, with brown markings; a so-called hermaphrodite specimen was obtained by Mr. Saville Kent ·in which the two sides displayed respectively the two types of coloration. One of the parrot-fishes also, *Pseudoscarus rivulatus*, displays marked sexual coloration, the female being blue and yellow and the male green and red, the prevailing tints during life being respectively blue and green—an interesting case because it recalls the conditions seen in some parrots. In connection with the coloration of the parrot-fishes there has been noticed a case of so-called "mimetic" resemblance in the case of a goby (*Gobius douglasi*), which is green banded with red, the usual colours of the parrot-fish. The giant anemones already mentioned have frequently small commensal fishes living in their central cavities; thus *Discosoma kenti* is inhabited by *Amphiprion percula*, a small red fish with broad white bands, the bands being separated from the red ground colour by a black margin; *Discosoma haddoni*, on the other hand, is inhabited by *Amphiprion bicinctus*, which is closely similar in colouring but is without the black margins to the white bands. Both sea-anemones show great colour variation, but in all cases their colours are sharply contrasted with those of their commensals.

It is interesting to note that the brilliancy of tint is entirely confined to the fishes which actually live among the brightly coloured corals. Those living in the lagoons which are floored with coral sand and have a scant fauna and flora are all dull of tint. This has been explained on the usual assumption of protective coloration, but it seems difficult to see that any explanation is necessary beyond that of the simple fact that in those situations the conditions of life are relatively unfavourable, that therefore dominant or brightly coloured forms are not likely to occur.

2. *Deep-sea Fishes.*—With the colours of these reef-fishes we may contrast the type of coloration seen in deep-sea forms. These are usually remarkable for their uniformity of colouring, bands, spots, or stripes being rare. Dark brown or black colours seem on the whole to predominate, and it is not infrequent for the mouth and gill cavities to be very darkly pigmented with black. After brown and black colours come violet or yellowish tints, usually dull-coloured, although a few examples are known of deep-sea fishes showing considerable brilliancy of tint.

3. *Temperate Fishes.* — It is unnecessary to describe the colours of the fishes found at moderate depths in temperate waters. The silvery iridescence of the lower surfaces of many, the frequency of dark spots or stripes, the not uncommon occurrence of red or orange pigments in the skin or muscle are familiar to all.

In general, fishes display considerable beauty of colour, but usually of a somewhat fugitive type. Under the most favourable conditions the beauty is in large part lost during any process of preparing the skins for mounting ; and it is probably to this rather than to any other cause that the prevalent impression that fishes are dull-coloured as compared with birds is due. Sexual dimorphism of colour is common both in those from temperate and from tropical seas, and so-called protective coloration is not infrequent.

The foregoing remarks, it should be noticed, refer almost exclusively to the modern dominant bony fishes or Teleosteans. Sharks and rays, as typical of the older Elasmobranchs, are dull in colour without the silvery sheen of modern fishes ; and although they may display a certain amount of beauty of marking, the colours are relatively dull and sombre.

The colours of fishes are due to the structure of the dermis or to colouring-matter contained in it. The colouring-matter is in part at least deposited in contractile pigment cells—chromatophores—the result being that many fishes, like Amphibians, are capable of a considerable amount of colour change, varying according to a familiar observation with the colour of the ground upon which they lie. The mechanism is set in motion through the eye, the phenomenon not being observable in blind fishes. Mr. Cunningham (1893, b) states that in the case of the flounder (*Pleuronectes*) deficiency of oxygen or alarm causes the chromatophores to contract, and so diminishes the intensity of the colour ; exclusion of light causes them to expand and so deepens the colour. Probably, as in the case of the chameleon, the colour depends in part upon the psychological state of the individual. Direct mechanical stimulation also causes the pigment cells to contract.

COLORATION OF FLAT-FISHES

It is impossible to speak of the coloration of fishes without touching, however briefly, on the classical problems connected with the coloration of flat-fishes. As is well known, these fishes are much-modified forms, being flattened from side to side, and having the upper surface, that is one of the sides, coloured usually in relatively dull tints, while the actual lower surface, strictly the other side, is of a silvery white colour without pigment. According to many naturalists, the colours of the upper surface are protective, the lower surface is without pigment, because it is usually concealed in these ground fishes, and therefore colour, if present, would be useless. Others, notably Mr. Cunningham, regard this explanation as totally inadequate, and hold that the absence of colour on the lower surface is due directly to the absence of light, or rather was so due in the first instance. To use Mr. Cunningham's own words (1893, a) : " The disappearance of the pigment from the lower side in the normal flat-fish is an hereditary character, not due to the withdrawal of the action of light in the individual. . . . On the other hand, the fact that the pigment, after prolonged action of the light, actually reappears, is strong evidence that originally, in the beginning of the evolution, the pigment disappeared, in consequence of the withdrawal of the lower sides from the action of the light. If so, an acquired character has become hereditary." We cannot here discuss the main question of the inheritance of acquired characters; it is unfortunately not one which is likely to be settled by direct experiment, but the careful observations to which the dispute has given rise are worthy of all attention.

In most Pleuronectidæ the upper surface is coloured with black and yellow pigment, the lower is pure silvery white; the peritoneum also is darkly pigmented on its upper surface and white on the lower, and this in spite of the fact that the upper part does not shine through the body walls at all and the lower very little. Two cases have been described, however, in which the lower surface of the skin is normally pigmented. In *Pleuronectes cynoglossus* the lower surface is gray and not white, the colour being due to the presence of black but not yellow pigment. This form "has been taken at all depths up to 700 fathoms." The upper surface contained both yellow and black pigment, but was not so darkly pigmented as in the case of shallow-water forms. It will be noted that this appears like an approximation to the uniform colouring of deep-sea forms. The other case is that of *Engyophrys sanctilaurentii*, in which the blind side has five or six dusky bands of colour occurring in the anterior half of the body; the posterior half is colourless.

Mr. Cunningham has made a series of careful experiments on the artificial production of pigment on the lower surface of flounders. The method adopted was to keep the young flounders in tanks under such conditions that the lower surface could be artificially illuminated by means of mirrors. The result of the experiments was to show that the amount of pigment developed increased with the amount of exposure, and that although pigment was developed more rapidly when the larvæ were very young, yet the power of developing pigment was not confined to any particular age. This is interesting as showing that the abnormal pigmentation was not a purely larval phenomenon as, for example, was the peculiar colour change shown by Fischel's salamanders (see Amphibians, p. 233). The illumination of the lower surface during the process of metamorphosis did not in any way interfere with the ordinary course of development, but throughout the whole of the experiments Mr. Cunningham noticed that the larvæ all had what he calls the "objectionable habit" of clinging to the sides of the tank so as to avoid as far as possible the illumination of the lower surface. This suggests that the light had some powerful and unpleasant influence upon the nervous system. The pigmentation of the lower surface was apparently in no case so marked as that of the normal upper surface, but was always the same in kind, and showed the same variations of colour due to contraction or expansion of the chromatophores as the upper surface. Artificially produced pigment always appeared first in the middle region of the body on each side of the lateral line, and last in the head and tail regions.

Besides these artificially produced abnormalities in flat-fish, many cases have been described in which the lower surface is more or less coloured under natural conditions. In some cases the colour is due to mere ill-defined patches of pigment, but in others the lower or blind surface shows in whole or

in part a type of coloration in all respects similar
to that of the upper surface, the similarity being
carried out into the details of spots and markings.
Such forms are called " double " or " ambicolorate."
This peculiarity has been described in the turbot
(*Rhombus maximus*), the brill (*R. lævis*), the flounder
(*Pleuronectes flesus*), the plaice (*P. platessa*), the
merry sole (*P. microcephalus*), and perhaps occurs
in others. An exceedingly interesting point about
the variation is that in the turbot, the brill, and the
flounder it is when marked always associated with
a peculiar malformation of the head, due to a growing forward of the dorsal fin. In the plaice it is
occasionally so associated, in the merry sole rarely
if ever. In the turbot the correlation is so exact
that, according to Mr. Cunningham, "if we draw an
imaginary line through the preopercular bone in the
turbot, pigmentation may extend over the whole
of the lower surface behind this line without any
structural malformation being present, but when
pigment is also present on the lower side in front of
this line the characteristic structural malformation
occurs also ; on the other hand, the structural malformation has never been observed in any specimens in which the lower side was unpigmented or
pigmented to a less extent than that defined above."
The meaning of the structural malformation is not
quite clear, though it has been suggested that it is
due to a delay in the shifting of the eye from the
blind side ; to explain its association with the pigmentation of the lower surface, it has been further
suggested that the delay allows a "continuation of
the power of receiving visual sensations from this

side" (Bateson). This hardly, however, explains the detailed correlation seen in the turbot, and the whole subject is still imperfectly understood.

As to the meaning of the similarity between the upper and lower surfaces in the ambicolorate specimens, both Mr. Bateson and Mr. Cunningham point out that this cannot be due to reversion, for there is no reason to suppose that the upper and lower surfaces of the ancestral Pleuronectidæ were identical in colour. Mr. Bateson regards it as a case of what he calls homœotic variation. "In the flatfish the right side and the left have been differentiated on different lines, as the several appendages of an arthropod have been, but on occasion the one may suddenly take up all or some of the characters, whether colour, tubercles, or otherwise, in the state to which they have been separately evolved in the other" (*Materials for the Study of Variation*).

In summarising the points which give their special interest to the coloration of flat-fish, we may note the usual absence of pigment from one side, right or left as the case may be, the occurrence of two forms which normally possess traces of pigment on this side, the facility with which pigmentation of the lower surface may be produced by artificial illumination, the occurrence of ambicolorate forms as a frequent variation in nature, and finally, the extreme difficulty of accounting for this variation on any hypothesis of reversion.

PIGMENTS AND STRUCTURAL COLOURS OF FISHES

As to the detailed characters of the coloration of fishes, there are several points of great interest. We have already mentioned the occurrence of crystals of guanin in the skin, which are important factors in the production of the iridescence and the silvery appearance of many fishes. The exact effect of these crystals on the coloration has not, however, been sufficiently determined, and it remains uncertain how far they are instrumental in producing the gorgeous colours of many tropical fishes. There is every reason to believe that these are optical in nature, but exact investigations into the mechanism of production are still required. Owing to the fugitive nature of the colours, the investigation would need to be conducted on the spot where the fishes are found, so we must look to some tropical biological station of the future for the complete solution of the question.

The most important recent work on the colours of fishes is that of Mr. Cunningham and Dr. M'Munn, who investigated especially the colours of the Pleuronectidæ or Flat-fishes. The first point is to realise the position of the coloured structures. We have already noticed that fishes possess some power of adapting their colours to their surroundings, and this proves at once that the colouring-matter is deposited in living cells and is not a cuticular product. In fact, it is the dermis which contains the elements giving rise to the coloration. It will be remembered that the scales of fishes are, for the most part, dermic

outgrowths, and that in fishes in general the epidermis
is little specialised ; in accordance with this we find
that the epidermis is usually almost entirely devoid
of pigment, or when, as in the flounder (*Pleuronectes
flesus*), some pigment is present, it has no effect on
the visible coloration.   In birds, on the other hand,
where the epidermis gives rise to greatly specialised
outgrowths, the feathers, the pigment of the body
is found in these epidermal structures.   There is,
however, a general consensus of opinion that in all
cases the pigment originates either in the dermis or
in deeper connective tissue cells and migrates outwards to the epidermis, in the cases where it is
found there.   Many would regard this as a proof
that pigments are in essence waste products, and
that this migration outwards is to be interpreted as
a process of excretion, the connective tissue cells or
amœbocytes carrying the waste pigment from the
essential organs of the body outward to the inert
skin.   For an elaborate hypothesis of this kind the
reader may be referred to a paper by Mr. H. E.
Durham ; it is doubtful, however, whether such
generalisations are not as yet premature.

To return from this digression to the immediate
subject of the skins of fishes, we find that, according to Cunningham and M'Munn, the elements important in coloration occur in two layers in the skin,
and the two layers differ considerably in the upper
and lower surfaces of the fish.   In most cases the
outer layer occurs in relation to the scales, while the
inner layer lies close to the surface of the muscles,
but in the mackerel the loose scales contain no
colouring elements.   The outer layer consists of

often branched and contractile chromatophores, containing different pigments, and of polygonal cells of peculiarly silvery appearance called by Pouchet iridocytes. The silvery iridescent appearance is due to the presence of crystals of guanin. The inner layer consists of the same elements. On the lower surface of the skin the outer layer contains iridocytes but no chromatophores, the inner consists of an argenteum—that dead-white opaque layer which clings to the surface of the muscles and is familiar to all who have examined fishes. Though no histological elements can be found in the argenteum in the adult, development shows, as might be expected, that it is formed from the fusion of iridocytes. The silvery or dead-white appearance is again due to crystals of guanin. The distribution of the colouring elements of the skin may be summarised in the following table :—

	Upper Surface.	Lower Surface.
Outer layer	Iridocytes and Chromatophores	Iridocytes only
Inner layer	Iridocytes and Chromatophores	Argenteum

This table is true for the flounder and, *mutato mutandis*, for other fishes as well. The points of importance are the presence of a large amount of reflecting tissue, especially on the lower surface, and the absence of pigment cells on that surface. The relative development of the parts varies of course greatly in different fishes, but the distribution appears to be fairly constant. The two layers of reflecting tissue seem to have different functions in some cases at least, for while the outer layer may exhibit brilliant iridescence, the inner " presents either a chalk-white opaque surface or an evenly bright silvery surface."

The pigments contained in the chromatophores, so far as they are at present investigated, appear to be always either lipochromes or dark melanins. The lipochromes are apparently very widely distributed and exhibit many different shades of colour. They are chiefly known by their spectroscopic characters.   In some cases, as in many of the flat-fishes, the yellow and black pigment cells produce a direct effect upon the coloration—that is, the combination of the two pigments produces a brownish colour tending either towards yellow or black, according as the one or the other pigment is most abundant.   Similarly in the gurnards some species like *Trigla lyra* are red, and others like *T. gurnardus* are gray, the difference being, according to Cunningham and M'Munn, merely due to the relative development of black and red pigment.   There are, however, other cases of greater difficulty in which the colour is not obviously due to the combination of two pigments.   Thus the beautiful green colour of the mackerel is produced not by a green pigment, but apparently by a blending of black and yellow chromatophores.   In the black bands the black chromatophores are much more numerous than the yellow, while in the green bands the two kinds are equally abundant.   The statement made above, that the green is the result of the " blending " of the two colours, is due to Mr. Cunningham and Dr. M'Munn, but it is a little difficult to believe that there may not be in addition some assisting optical effect. A similar but even more interesting case is that of the green pipe-fish (*Siphonostoma typhle*), often quoted as an instance of protective resemblance. In this form a similar combination of yellow and black pigment occurs, but there is some doubt whether the green colour is not in part due to the action of the reflecting tissue. This pipe - fish, though usually green, may appear in a brown form, while the common pipe - fish (*Syngnathus acus*) is always brown. The common pipe-fish also contains yellow and black pigments, but the yellow is said by the authors to be more orange-coloured than that of the preceding form. It is, however, curious to note that the three pigments, that of the mackerel (*Scomber scomber*) and those of the two pipe-fishes, yield solutions whose spectroscopic characters are almost the same; this certainly suggests that the green is primarily a result rather of structure than of pigment. True green pigments in fishes were, however, described some time ago by Mr. G. Francis in certain of the Wrasses (species of *Odax* and *Labrichthys*), but the observations have never been repeated. The Mediterranean members of the family are said by Krukenberg to owe their green and blue colours to structural effects.

In general, we may say of the colours of fishes that they are either pigmental or structural, and that the only pigments which have been described with certainty are lipochromes or melanins. In most cases both kinds of pigments occur simultaneously, but in some, as in the gold-fish (*Carassius auratus*), the melanin pigments are entirely absent. The presence of a green colour is frequently associated with a combination of yellow and black pigments, but the green colour is probably always in part produced by the structure of the tissues. The silvery whiteness of the lower surface of most fish, and the gleaming iridescence of the other parts of the body in some, are both due to the abundant deposits of guanin in the skin, but the relation of this reflecting tissue to the brilliant evanescent colours of many tropical fishes is unknown. Spots or bands produced by the irregular distribution of the colouring elements are common, but the reason of this irregularity of distribution is quite unknown. Usually the lateral line is marked externally by a distinct band of pigment, but in some cases, as in *Atherina presbyter*, it is marked instead by a white band due to a special development of reflecting tissue (Cunningham).

The colours of the lipochrome pigments vary from yellow to red, but the yellows are apparently the most common. In some instances the skin is sufficiently transparent to allow the subjacent muscles to shine through, and these may be so coloured with hæmoglobin as to have a distinct effect upon the coloration.

According to Mr. Beddard (*Animal Coloration*, p. 11), the green colour of the bones of the fishes *Belone*, *Protopterus*, and *Lepidosiren* is due to the presence of the mineral vivianite, but no reference is given.

CHAPTER XI

THE COLOURS OF AMPHIBIANS AND REPTILES

Colours and Pigments of Amphibia—The Colours of the
Larvæ—Development of Colour—The Relation of the
Larval and Adult Coloration—Colour Variation in Larvæ
—Colours of Reptiles, especially Lizards and Snakes—
The Origin of Markings in Snakes.

IN Amphibia the colours are not infrequently sober
and incline towards the so-called protective tints, but
in some cases they are bright and conspicuous. The
skin is smooth, soft, and scaleless, and the power of
colour-change not infrequently well marked. As a
whole, the prevailing colours are dull brown or
black, shades of green, or vivid reds and yellows,
thrown up against a dark background. As examples of forms showing dull coloration we may
mention the Mexican axolotl and the common
newt. A green colour is well shown by the tree-
frogs (*Hyla*), often kept as pets in ferneries. The
spotted salamander (*Salamandra maculata*), with its
vivid colouring in black and yellow, and Mr. Belt's
famous blue and red frog show, however, that more
brilliant types of colouring are far from being absent

Q

in Amphibians. The power of colour-change is familiar in the case of the common frog. A very little experimentation will show that in it the general colour is dull and dark against a background of earth or peat, and bright yellow-green among fresh herbage. In spite, however, of this power of colour-change, and of individual colour variation, there is much constancy of marking. In our common British frog there are two longitudinal stripes running down the sides of the body which seem to be absolutely constant, and which appear in the larva at the commencement of the metamorphosis. All those who have kept tadpoles in confinement must have noticed this fact, and learnt to regard it as a sign that their pets will shortly require a complete change of environment— from water to land. Markings of this kind occur constantly both in Amphibians and Reptiles, and are of much importance. Their origin has been investigated in one case only, in a snake, and it has been found that, as might be expected, they are closely related to internal structures. Further investigations on the same lines would be of much interest. That they are constant throughout large groups and dependent upon "laws of growth" has long been maintained upon theoretical grounds by Professor Eimer and his school.

As to the details of the mechanism of colour, the slight development of the epidermis, and the power of colour-change, show at once that the elements important in coloration must occur in the true skin or dermis. The epidermis does contain a small amount of pigment granules, but these are unimportant as compared with the contractile pigment cells of the dermis. The colour-changes are due to variations in the degree of contraction and expansion of these pigment cells, or of their protoplasmic contents.

The pigments, where they have been investigated, have been found, as in the case of fishes, to be either lipochromes or melanins. Guanin is also present in the skins of Amphibians, but the amount is smaller than in fish, and the effect on coloration much less obvious. The lipochromes and melanins, occurring separately or combined, produce such colours as black, brown, yellow, orange, and so forth, while in association with optical effects they produce such colours as blue and green. The skin of the frog when steeped in alcohol loses its green colour owing to the fact that the lipochrome dissolves out. It is, however, an old observation that the colour will return if the grayish skin be covered over with wet yellow tissue paper (Krukenberg).

THE COLOURS OF THE LARVÆ

The Amphibia display in general a very marked metamorphosis during the course of development, and there are often notable differences in the colour of the larval and adult stages. The larvæ are peculiarly susceptible to environmental influences as regards colour, but they differ from the adults in apparently being more sensitive to heat than to light, and also in the relative stability and permanence of the effect produced. There is some evidence to show that the elements producing coloration are more uniformly distributed in the larvæ than in the adults, and that development is accompanied by a re-distribution of pigment as well as by changes in amount.   It is interesting to note that in Amphibians as in Fishes colour-changes, whether natural or artificially produced, are not confined to the skin but occur also in internal structures, *e.g.* the peritoneum.

In studying the pigmentation of larvæ we may take first the case of the spotted salamander (*Salamandra maculata*), where the coloration has been described in detail by Fischel.   This salamander is often kept in confinement, and the vivid black and yellow colouring of the adult is very familiar.   The larvæ under normal conditions when nearly full-grown are almost entirely black, the black pigment occurring in spots on a somewhat lighter ground.   A histological examination shows that the colour is due to the action of four factors, which are as follows :—

(1) Pigment granules lying 'in the cells of the epidermis.

(2) Branched pigment cells lying between the epidermal cells, sending their processes into the intercellular spaces.

(3) Similar branched pigment cells lying beneath the epidermis in the true skin or dermis.

(4) Pale yellow cells similar in character and position to (3) but in the normal condition largely concealed by these.

In the first three cases, the pigment is melanin of dark brown or black colour, in the fourth it is yellow lipochrome.   The pigment cells of the dermis— black and yellow—are by far the most important factors in coloration, the epidermic pigment having relatively little effect. In the larva, the dermis over its whole area contains both black and yellow cells, but the latter are more abundant at certain regions, corresponding to the future yellow spots of the adult. The peritoneum contains also both yellow and black cells, while in the adult there is no trace of yellow cells in it.

Fischel's observations on the coloration were made in connection with some experiments on artificially produced colour-change, and he did not therefore carry them beyond this point. Before proceeding, however, to consider the theoretical bearing of the facts given above, we may supplement them by some notes made by Dr. Bedriaga on the development of colour in the larvæ of newts. Bedriaga did not investigate the histology of his specimens, but in view of the similarity of colour, we are probably justified in applying Fischel's statements throughout.

We may take the case of *Molge montana*, which is described in some detail, as typical of one set. In this form, larvæ 10-15 mm. in length are uniformly yellow beneath, and yellowish-white with black spots above. Larvæ 20-25 mm. in length show these black spots spreading over the light ground, especially on the upper surface. From 25 up to 40 mm. the larvæ show a gradual increase of the black pigment which now becomes the ground colour, the light pigment appearing in the shape of small spots, but forming also a median yellow stripe. Larvæ of 40-45 mm. length appear when taken from deep water to be perfectly black, but if placed in shallow vessels they become much lighter, and then exhibit very numerous black spots on a dull gray or yellow-gray ground. This type, in which the very young larvæ have dark spots on a pale ground on the upper surface and a pale unspotted lower surface, is exceedingly common among species of *Molge*. During the course of development the upper surface tends to darken until it presents the appearance of a few light spots on a dark ground, while the primitively uniform lower surface becomes spotted with dark colour. Development, therefore, seems to be accompanied by an increase of dark pigment, and a decrease of yellow or lipochrome pigment. This type, though common, is not universal among the species of *Molge*; some, like the common frog, are dark at first and gradually grow lighter. Thus very young larvæ of *Molge alpestris* occur in two varieties. In the one the upper surface is covered with dark-brown spots closely connected together so as to form a network whose meshes are pure brown, while the lower surface is clear and unspotted. The other variety is similar except that the meshes of the network are grayish-yellow instead of brown. As the larvæ increase in size, the meshes of the network grow larger and paler in colour until they become grayish-brown or yellow, the network itself at the same time growing paler in colour. Throughout, however, the distinction between pale and dark larvæ is preserved, and there is evidence to show that the dark are the future males, and the light the future females.

The larvæ of *Molge alpestris* like many other larval Amphibians may become sexually mature before metamorphosis. In that case the colours differ slightly from those of the normal adult, the difference expressing itself especially in diminished brightness of tint and in the absence of the cuticular modification which gives rise to a blue colour.

RELATION OF LARVAL AND ADULT COLORATION

It is unfortunate that the paucity of observations makes us unable to say much on this interesting subject, but the statements given above suggest one or two points.

First as to the position of the pigment and its meaning. We have seen that the dark pigment, both in adult and in larva, occurs to a slight extent in the epidermis, and much more markedly in special pigment cells in the dermis or true skin. The epidermis, with its contained pigment, is periodically cast and renewed, so that there must be a slow elimination of pigment from the body in this way. We have already seen that it is most probable that the pigment originates in the dermis or deeper tissues, so that there must be a continual migration of pigment from the dermis to the epidermis. Such a state of affairs occurs very frequently in vertebrates, and is, as noted above, often regarded as a proof that the coloration is the result of an excretory process. The development of the coloration, as already described, supports this theory in so far as it tends to show that the dark pigment usually increases in amount during growth, and that it tends to be more abundant in males than in females, in normal individuals than in those which are precociously sexual. The proof would, however, be more cogent were it not for the presence of the lipochromes. These can hardly be regarded as of excretory nature, they apparently tend to decrease during the process of development, but, on the other hand, they are abundant in adults showing specialised colouring, as in the case of the spotted salamander, and often increase in amount during the breeding season in individuals which exhibit a seasonal change. Reinke (*Arch. f. mikrosk. Anat.* Bd. 43, 1894) has sought to solve some of those difficulties, and especially the disappearance of the yellow cells from the peritoneum, by the supposition that the yellow pigment may be converted into the black. This is strongly opposed by Fischel, and the question apparently does not at present admit of decisive settlement. It thus seems that the simultaneous existence of both black and yellow pigments prevents us here, as in so many cases, from accepting in its entirety the theory that pigmentation is the result of an excretory process.

Although our present knowledge does not enable us to make any statement as to the origin of the lipochromes in Amphibia, yet the developments described above disclose some interesting facts in connection with their distribution. The statements of Fischel and Bedriaga taken together show clearly that in the early stages of development the two sets of pigments are closely intermixed, and that the growth of the larvæ or the transition from larval to adult life, tends to be accompanied by a process of segregation, the yellow pigments accumulating at certain spots and the black at others. The process is an interesting one because it seems to occur frequently in birds, in the males as compared with the females. Its external result is to greatly increase the purity and brilliancy of the colouring, as well as to give rise to markings. An inquiry into the causation of the process would be of great interest.

COLOUR VARIATION IN LARVÆ

As we have already seen, the larvæ of Amphibians are very sensitive to environmental influences, their colour changing according to the conditions under which they are found. This fact was noticed both by Fischel and Bedriaga, and the former made some extensive experiments on the subject. He and one of his colleagues obtained a number of larvæ of *Salamandra maculata*. Half were put in a fish-hatching apparatus, and the other half in standing water in a dish. The former displayed the ordinary type of coloration, the latter were distinctly lighter without displaying any other abnormality. The ground colour was a light yellowish-white, and the ordinary black spots were only represented by slight traces of dark pigment, occurring especially at the posterior end of the body. As well as displaying little pigment, the larvæ are said to have had a peculiar glassy and transparent appearance. The larvæ were not all uniform, but showed considerable variation, those in the standing water were, however, all much lighter than those in the running water.

On making a histological examination of the skin of the pale larvæ, it was found that the cells of the epidermis contained a much smaller amount of pigment than usual, while the branched pigment cells of this layer and of the underlying dermis were represented by dark-coloured oval or rounded bodies.

A careful examination of a series of specimens showed that these structures represented the normal branched pigment cells, whose processes are here completely contracted. The yellow cells of the dermis were, however, not contracted, and now displayed themselves as a delicate network of branched pigment cells ; it is to these that the pale larvæ owe their colour. The difference in tint between the two sets of larvæ is thus ultimately due to the differing susceptibilities of the pigment cells of the skin. The peritoneum exhibited the same variation in colour as did the skin.

As to the cause of the difference, experiment convinced Fischel that it was due to the difference in temperature of the water in which the larvæ were reared ; that of the standing water being 9° to 11° higher than the running water. This was confirmed by removing dark larvæ from the cold water and placing them in warm water, when it was found that in the course of a day or two they had become notably lighter, and at the end of at most a fortnight were completely converted into the pale form. Although the pale larvæ so produced were quickly reconverted into the dark form, it was not found so easy to render the original light forms dark. A long exposure to the influence of warm water apparently renders it difficult for the pigment cells to become fully expanded when exposed to cold, no doubt also these larvæ were actually deficient in dark pigment. The colour-change could only be produced in young larvæ, older ones having apparently lost the power of reacting to heat or cold. The effect of prolonged exposure to heat or cold is, therefore, to produce a stable unalterable type of coloration.

The effect of light on the developing larvæ was also tested, and it was found that light has but little effect, though on the whole the larvæ are darker in light than in darkness. This is interesting because the same thing has been noticed in tadpoles of the frog, which become pale and transparent in darkness, while adult frogs, many fishes, reptiles, etc., become pale in strong light. Thus at least in the frog the larva and the adult react differently to the stimulus of light.

As an interesting commentary upon these observations of Fischel's, we may note that Bedriaga in describing the larvæ of *Molge aspersa*, remarks that while the colour is usually an olive-brown with yellowish spots, larvæ taken from deep water are much darker in tint, and the yellow spots are reduced both in size and intensity. On the other hand, larvæ from sunny shallow situations are pale in colour and very distinctly spotted with yellow. As Fischel found that light as a whole tended to darken the colours, we must suppose that in this case it is the temperature of the water in the two situations which is important in affecting coloration.

Similarly, Karl Knauthe found that adult Amphibians (*Bufo variabilis* and *B. vulgaris*, *Pelobates fuscus*, and others) turned dark, often gray or black, when exposed to the influence of intense cold. It would appear from this observation that the larvæ and adults react similarly to variations in temperature.

Colours of Reptiles, especially Lizards and Snakes

Among Reptiles the colours are most conspicuous, and have been most studied among snakes and lizards.   In lizards black, gray, brown, and green are common colours, while most display very distinct and constant markings.   According to Eimer the typical ground plan of the dark markings consists of seven longitudinal bands.   The pigments seem to consist of the dark "melanins" and lipochromes, the green colour of, *e.g.*, the green lizard being not due to a green pigment but to the structure of the skin, and to the contained yellow and black pigments. According to Krukenberg lipochromes are rare or perhaps absent in snakes, while melanins largely predominate.   Yellow pigments do occur freely in snakes but their nature is undetermined, probably they are merely altered lipochromes ; for the peculiarities which Krukenberg enumerates may be due to impurities or to decomposition.   Lizards frequently exhibit the type of coloration known as protective, and, as is well known, have often considerable power of colour-change, best exemplified in the chameleon. It is interesting to note that the aberrant poisonous lizard *Heloderma* diverges markedly in colour from other lizards, being vividly marked with black and yellow, instead of showing the sober greenish hues of most lizards with their beautiful but unobtrusive striping.   In its histological characters the skin of lizards shows much similarity to that of the frog, the power of colour-change, when present, being again due to the varying susceptibilities of the chromatophores to the action of light. For an elaborate discussion of the coloration of the wall-lizard (*Lacerta muralis*) the reader may be referred to Eimer's well-known papers.

The colours of snakes are often dull, and due in large part to obscure mottlings, but in some cases, as in the deadly coral snake (*Elaps*), there is a conspicuous black and red banding. As compared with lizards, snakes seem to be characterised by an increasing predominance of epidermic pigments which are got rid of and renewed when the slough is cast. Many would regard this again as a proof that the pigments are waste products, in process of being eliminated by means of the skin. The skin of snakes, like that of most reptiles, contains guanin, which is said by Leydig to give rise in some cases to white and yellowish patches.

Origin of Markings in Snakes

In discussing the coloration of leeches we have seen that an attempt has been made to explain this on mechanical grounds, by associating the stripes or spots with the development and arrangement of the muscles—it is interesting to note that in snakes the coloration seems to bear a similar relation to the arrangement of the blood-vessels. Herr Jonathan Zenneck finds that in the case of the ringed snake (*Trophidonotus natrix*) the three longitudinal rows of spots in the adult correspond to three red lines in the embryo which mark the course of subcutaneous blood-vessels. Zenneck found that in the adult the black spots were produced by accumulations of black pigment in the cutis of these regions, in other respects the skin showed no special peculiarities. He therefore set himself the problem to find out what anatomical or developmental necessities produced these aggregations of pigment.

The ringed snake is marked by three paired longitudinal rows of black spots, which correspond to the longitudinal lines described by Eimer in the wall-lizard, except that in the lizard there is, in addition, an unpaired median line on the dorsal surface. In the embryo at an early stage the surface is marked by three pairs of red lines, and a slender median unpaired one. The lines are produced by the subjacent blood-vessels and are not necessarily perfectly continuous, being sometimes broken into red spots ; they are connected with numerous transverse vessels, the junctions being marked by distinct swellings of bright colour. Of the three lateral lines the median is the most distinct, and extends forward in front of the eye. A little later the scales begin to develop from before backwards, and as they become distinct the red lines disappear before them, and, at the same time the first traces of pigmentation appear as a longitudinal row of pigment spots at each side, developed in the position previously occupied by the median red line already mentioned. The extension of this line in front of the eye is now marked by a distinct spot of pigment, which forms a marked feature in the adult. The formation of this middle row of spots is followed a little later by the appearance of two other rows corresponding to the upper and lower of the lateral red lines. The first spot of the middle row at a later stage fuses with the first of the upper row at each side, and the two large spots lying close together form the black part of the "collar" of the adult. The author does not state what becomes of the unpaired median red line, it presumably disappears without being replaced by pigment spots, but its presence is interesting in view of the fact that the lizard has a black line in this region.

An examination of the red lines, by means of sections, shows that they are due to superficial blood-vessels connected at regular intervals with the deeper vessels of the body. The median red line of the three laterals corresponds to a vein called the epigastric which receives blood brought back from the vessels of the skin, and transmits it *viâ* a series of transverse vessels to the deeper veins, *e.g.* the cardinal vein. The first appearance of pigment, except in the choroid, is in connective tissue cells lining the body cavity, and these appear to spread round about the epigastric, though this is uncertain. At any rate the epigastric vein loses its endothelial lining, becomes filled with connective tissue cells, and is gradually obliterated, the obliteration being preceded by a formation of direct connections between the vessels of the skin and the transverse vessels, so that the epigastric vein is thrown out of the circulation. At this period the pigment spots are developed, and they occur in the cutis opposite the points where the transverse vessels formerly entered the epigastric. A similar relation exists between the other rows of spots and blood-vessels.

Herr Zenneck leaves undecided the question whether the pigment originates entirely in the connective tissue cells of the body cavity, and migrates outwards through the blood-vessels in the wandering cells which fill up the cavities of these, or whether it may arise in part in the epidermis *in situ*, and confines himself to emphasising the fact of the relation between the pigment spots and the obliterated openings of the transverse vessels. When we remember, however, that the places where blood stagnates are especially liable to become deeply pigmented, it is difficult to avoid the conclusion that the pigment directly originates from the degenerating blood which must accumulate at these points. The question has some bearing upon theories of the origin of colour and colour patterns, but is obscured by the usual difficulty of distinguishing between *post* and *propter*.

In connection with this paper of Zenneck's we may note an observation by Mr. J. Loeb on the relation between the blood-vessels and the coloration of the yolk-sac in the embryos of the fish *Fundulus*. The yolk-sac has here a peculiarly tiger-like colour due to a combination of black and red chromatophores. At the time of their first appearance the pigment cells are practically uniformly scattered, but as the blood-vessels develop and the blood begins to circulate, the chromatophores begin to migrate to the surface of the vessels, and ultimately pigment is only visible as a covering to the vessels. By an ingenious experiment with a heart poison (potassium chloride), Loeb convinced himself that the migration only occurred when the blood was actually circulating in the vessels of the embryo. He concludes that the coloration of the yolk-sac in *Fundulus* is due to a specific irritability of the

chromatophores, comparable to the irritability displayed by many leucocytes, which forces the chromatophores to migrate to the surface of the blood-
vessels. This observation affords an interesting
parallel to those of Zenneck's, except that in the
latter case the chromatophores congregated about
degenerating blood-vessels, and in Loeb's case about
living vessels. Loeb considers that the typical
coloration of the embryo is similarly related to the
distribution of its blood-vessels. He promises
further observations on the subject.

List describes the pigment of the embryos of
bony fishes as originating from the debris of the
yolk. It is then taken up by wandering cells and
carried to different tissues. The relation between
blood-vessels and pigment cells described by Loeb
for *Fundulus*, List describes as universal for vertebrates. He considers that the surface of the blood-
vessels forms the main track outwards for the
pigmented leucocytes, which tend continually to
migrate from the deeper tissues to the epidermis.
The two investigators do not appear to have been
aware of each other's work.

The interest of these three sets of observations
is that they all correlate the production of pigment
and the development of markings with the physiology of the developing embryo, and suggest that
we may yet be able to similarly explain colour-
phenomena in general.

R

CHAPTER XII

THE COLOURS OF BIRDS

General Characters of the Coloration of Birds—Sexual Coloration—Distribution of Colour and Colour Variation—Food
and Colour—Pigments of Birds, their Characters and
Distribution — Pigments of Birds' Eggs — Markings of
Feathers.

THE colours of birds often rival or surpass those of
insects in brilliancy and variety, and have attracted
quite as much attention. Indeed, the greater size of
birds and their greater specialisation give to their
colouring a variety and a charm which many fail to
find in insects; this widespread admiration has unfortunately been in many cases singularly fatal to
the birds.

Birds resemble insects in displaying both types
of colour—the optical and the pigmental—to great
perfection; but in marked contrast to the conditions
which prevail in the Lepidoptera, we find that the
bright pigments are usually lipochromes, never so
far as is known waste products of the uric acid
group. Further, although there are several instances
described of birds whose colours can be heightened
or altered by the employment of special kinds of

food, there is at present no reason to doubt that under ordinary circumstances the lipochromes of birds are self-produced and not derived.

Let us first consider the distribution of colour in birds. The bright colours are largely, though by no means exclusively, confined to the exposed parts of the feathers. They may, however, occur on the beak, the feet, and legs, the bare patches on the head and neck of many birds, and even the parts of the body which are covered with feathers, and the mouth cavity. The colours in general fall into three sets :—Those due to lipochromes, those due to the dark melanins, and those which are structural. The lipochromes when present tend to be uniformly diffused, and are probably always in solution in the abundant fats. It is an old observation that the intensity of the red colour in the flamingo depends upon the amount of oil contained in the feathers. Besides occurring in the feathers, bill, feet, etc., lipochromes usually colour the deposits of fat in the body and the yolk of the eggs.

The dark melanin pigments are very widely distributed in birds as in all Vertebrates. They occur in the form of minute amorphous granules in the epidermis or the cuticular structures, and not infrequently give rise to brilliant structural colour or to very elaborate patterns and markings ; in some cases they are uniformly distributed and give rise to plain gray, brown, black, and related tints. Structural colours in birds are abundant, and include, in addition to all metallic colours, blue, green, some yellows, white, and in part the glossy black colours. Blue, whether it occurs on feathers as in the jay, or on the beak, or the skin of the head, as in some birds of paradise, is always structural. The naked patches of skin in birds indeed exhibit the same tendencies with regard to colouring as are visible in the feathers.

SEXUAL COLORATION

Before giving a detailed account of the colours and colouring-matters of birds we may summarise some of the general characters of the coloration. One of the most marked characters of the group is the great prevalence of sexual differences in colour, usually though not invariably of such nature that the male excels the female in brilliancy, or at least in intensity of colour. The colour differences are of many kinds. Thus, as in the humming-birds and the peacock, the male may display brilliant struc-tural colour, absent or feebly developed in the female ; or as in some of the birds of paradise, the plumage of the male may be coloured by special pigments which are absent in the female. Again, as in our own blackbird, the male may, as compared with the female, merely exhibit a greater intensity of colour. The same fact is illustrated in a curious way in the case of the satin bower bird (*Ptilono-rhynchus violaceus*) ; here the female has a grayish somewhat thrush-like plumage exhibiting through-out faint but distinct metallic tints, so that the whole bird has a delicately iridescent appearance. The male, on the other hand, is a deep glossy black, with a hard metallic lustre—a more specialised, if rather less beautiful colour. Another very interesting case is that of the genus *Pericrocotus*, in which, according to Professor Newton, the males are generally black and rose-colour and the females gray and saffron; this is probably again due to increased amount of pigment in the male. So far the sexual differences we have noted have depended on the development of structural colour in the male, on the development of new pigment, or on the increased amount of existing pigments. There is, however, another difference often marked, and that is the relatively greater purity of tint in the male, and the frequent presence of contrasting colours. Thus in the male blackbird the glossy plumage contrasts sharply with the bright yellow beak. In *Sericulus melinus*, one of the regent birds of Australia, the female is dull grayish-brown and speckled, while the male is black, with brilliant patches of bright orange. In the beautiful orioles (*Icterus*) of North America the females are olive-green, the males black and yellow; the true orioles of the genus *Oriolus* show the same sexual difference even more distinctly. In the North American jays the colours of the males are frequently blue, white, and black, with bars and spots, while the females and some unspecialised species are gray. Facts of this kind are of very common occurrence, and have been much insisted upon by Mr. Charles Keeler, who regards them as tending to prove that the general ground colour of the females or of unspecialised species is due to a mixture of pigments, while the separation of the pigments gives rise to the pure colours of the specialised males. The force of Mr. Keeler's arguments is diminished by the lack of precise discrimination between pigmental and optical colours; but the facts are at least interesting, and it is quite possible that the dull olive colours of many unspecialised birds may be due to a mixture of lipochrome and melanin, which, when separated, give rise to vivid orange and black colours in the males. We have already noticed facts of similar nature in larval Amphibians as compared with the adults (see p. 232).

DISTRIBUTION OF COLOUR IN GENERA AND COLOUR VARIATION

Closely associated with sexual colour differences is the question of the distribution of colour in the species of a genus, and in this connection a few examples of the relation between yellow and red are worth quoting from Mr. Keeler. That such a relation should exist at all is interesting, because it presents some sort of parallelism to the relation between red and yellow which exists in the Lepidoptera. In that group there is some reason to believe that the red pigment bears to the yellow a direct chemical relation, but the reds and yellows of birds are usually due to lipochromes, and the relation between the red and yellow lipochromes is still very obscure.

Mr. Keeler lays down the general rule that wherever red is present in a genus, yellow will also be present. Thus in the grosbeaks (*Habia*) the male of *H. ludoviciana* has a breast patch of bright red, some of the wing-coverts being of the same tint, in *H. melanocephala* these parts are lemon-yellow, while in the females of both species they are a pale yellow. The brilliant red of the scarlet tanager (*Pyranga erythromela*) is replaced in the western tanager to a large extent by yellow, while the females and young of both are yellow. So also in the American redstart (*Setophaga ruticilla*) the areas which in the female are yellow are orange-red in the male. In none of these cases do the lipochromes appear to have been investigated; it is most probable that in some at least the difference in colour is due merely to the amount of pigment present in the coloured parts, or it may be in part due to structural differences.

The question as to whether it is possible to speak of a geographical distribution of colour is an interesting one, but one which it does not appear as yet possible to decide. That tropical birds tend to be brilliant, and Arctic birds white, appear to be as yet almost the only certainties on the subject. White as a general ground colour is of considerable interest in birds; it is certainly most common among the birds of cold climates, but is there often very slowly acquired; the gannet, for example, takes several years to acquire the pure white adult plumage, and so with many others in which the adults are pure white. The existence of pure white birds is complicated by the frequent occurrence of albinos among many species normally coloured; "white" blackbirds are, for example, of common occurrence. This albinism may be complete, affecting even the eye, or it may be confined to special feathers or regions of the body. Now the natural whiteness of many birds is often compared to these cases of partial or complete albinism, although there is some doubt how far the comparison is legitimate. White patches are certainly often a sign of specialisation, and are not infrequently confined to the males. In the male of the king paradise bird (*Cicinnurus regius*) the ventral surface of the body is a pure white, adding greatly to the beauty of the coloration ; but the white feathers are a deep ashy-gray at their bases, and there can be little doubt that this is merely one of those cases of sifting of colour which are so characteristic of many male birds. The physiological meaning of cases like that of the gannet, however, where there seems to be a complete disappearance of colour, is very obscure ; to suggest that it is due to a constitutional change does not appear to advance the question much.

In connection with the subject of albinism it may be well to mention that the converse variation, that of excess of pigment, is said to occur both in the case of melanin and lipochromes, giving rise to melanism or erythrism. The fact that neither of these can be associated with the functional disabilities which are usually believed to result from total albinism puts them upon a somewhat different plane, but the whole question is difficult and doubtful. The same may be said of some other facts relating to colour, such as the tendency to melanism said to be exhibited by birds occurring in islands.

Another question of some interest is whether a change of colour can be effected in the plumage without a previous moult. This is strongly supported by some authorities and as strongly denied by others. Gätke, in his *Heligoland as an Ornitho-*

logical Observatory, gives numerous instances of this change occurring in birds, and distinguishes between cases in which the colour-change is associated with a change in feather-structure, and those in which it is not so associated. Thus in the linnet and mealy redpole, the surface of the barbs is said to peel off, exposing the fresh and bright colours beneath. In the case of the guillemot and little auks, on the other hand, he describes an increase in the amount of pigment in the feather without any textual change. Gätke enumerates numerous other instances of colour-change produced in this way by an increased amount of pigment, or by a rearrangement of pigment, as when black or blackish-brown replaces white or gray, or when black and white replace gray.

A change of colour accompanied by a shedding of part of the feather seems not inexplicable, but the mechanism of a colour-change without this is difficult to understand, and the fact has been strenuously denied by many. A recent summary by Schenkling gives an account of the various opinions which have been held on the subject. Schenkling himself strongly inclines against the view that a notable colour-change can occur without a moult or a shedding of portions of the feathers. He believes that the confusion has arisen from the fact that the same moult has not identical effects upon all the individuals of a species, or even upon all the feathers of an individual. It is thus possible to obtain specimens displaying feathers apparently characteristic of successive moults. Such specimens have been described as birds showing colour-change without moult, but it can be demonstrated that the feathers possess their peculiar colouring before they leave their sheaths. These statements are true only of birds which display complex colour-patterns, and which require years to completely attain the typical plumage. Schenkling is of opinion that a more careful study of such cases will greatly modify existing views as to the rigid limits of each moult.

FOOD AND COLOUR

It may perhaps be well to mention here for the sake of completeness the colour-changes which may be produced in some birds by supplying coloured food. An account of these changes will be found in Mr. Beddard's *Animal Coloration*. In the case of the canary it is well established that the addition of cayenne pepper to the food will change the colour from yellow to deep orange, or flame colour. Mr. Beddard also cites the artificially produced change from green to yellow in Brazilian parrots as another instance of the effect of food on colour, but there seems some doubt whether the change in this case is not produced by direct local application to the young feathers. In the case of the canary the change can only be produced in very young birds, which is so far evidence against the view that mature feathers can change colour. In the canary the change is produced by means of the intervention of a fat,—a point of some interest because the association of introduced pigments with fats is so common—is perhaps universal. Similarly the Rajah Lori is said to attain its brilliancy from a diet of

fish fat (cf. the statement as to the flamingo
above).

THE PIGMENTS OF BIRDS, THEIR CHARACTERS AND DISTRIBUTION

As to the details connected with the pigments of
birds, we may notice first that practically nothing is
known of the melanin pigments. They are widely,
perhaps universally, distributed in birds, give rise
to all the dull and sober tints, colour all feathers
displaying any beauty of marking, and are usually
associated with feathers displaying structural colour.
In birds as in mammals they form the groundwork
of the coloration. Of their characters little is known.
They may possibly be the same as the dark pigments of mammals; an origin from hæmoglobin has
been suggested here as elsewhere in Vertebrates, but
is not supported by very cogent evidence.

The lipochromes are numerous and diverse, and
have been chiefly investigated by Krukenberg.
The most familiar is probably the red pigment,
called by Bogdanow zoonerythrin, and by Wurm
tetronerythrin, which colours the red feathers of the
flamingo (*Phœnicopterus antiquorum*), of the cardinal
bird (*Cardinalis virginianus*), and of many others,
the red wattles in the male pheasant, etc. etc. The
only other red pigment certainly of lipochrome
nature which Krukenberg describes, is one which
he calls "araroth," found in the red feathers of
certain parrots, and differing only slightly from
zoonerythrin.

Of yellow lipochromes, on the other hand, he

describes a considerable number, differing from one another by their spectra, their solubilities, or their reaction to light. It does not appear that any good purpose would be served by giving a list of these pigments, at least in our present state of ignorance as to the relations of the lipochromes. It is sufficient to note that yellow lipochromes occur very frequently in the skin, the fat, the yolks of the eggs, and the feathers of birds; that these yellow lipochromes not infrequently occur mingled with zoonerythrin, that yellow feathers may contain two different yellow pigments, and that these yellow pigments may also be present in feathers which contain too much dark pigment for the yellow colour to be visible. The appearance of the feather is thus no certain criterion of the presence or absence of lipochrome pigment.

An exceedingly curious instance of this is afforded by Krukenberg's researches on the colouring-matters of *Eclectus polychlorus*. In this interesting parrot the male is chiefly green, with patches of red and blue; the female is chiefly red, with patches of yellow and blue; while, according to Meyer, the young of both sexes are red. The blue and green feathers are grayish-black on their lower surfaces and appear dull-coloured by transmitted light; the yellow and red do not change colour by transmitted light. The pigments present are one or more dark-coloured melanins, a yellow lipochrome, zoofulvin, and red "araroth"; the colour differences are in part due to structure, in part to the varying amounts of the pigments in the two sexes. This is shown in the following table :—

Male
{ Green feathers contain zoofulvin and melanin.
 Red   ,,   ,,   " araroth."
 Blue   ,,   ,,   melanin.

Female
{ Red   ,,   ,,   " araroth."
 Yellow   ,,   ,,   zoofulvin and " araroth."
 Blue   ,,   ,,   melanin.

It would appear from this that the melanin pigments are more abundant in the male, and the lipochrome in the female and the young. The case is a very interesting one, but it is doubtful how far it is safe to build conclusion upon it.

Besides the lipochrome and melanin pigments in birds, there are a few other isolated colouring-matters of some importance. Of these the best known is turacin, with which the name of Professor A. H. Church is so closely associated. Turacin is a reddish-purple pigment occurring in patches on some of the primary and secondary wing-quills of various of the Musophagidæ, or plantain-eaters, such as the type genus and *Corythaix*. The pigment is soluble in water, is said to be in part washed out of the feathers by heavy rain, and also to colour the water in which specimens kept in confinement are in the habit of bathing. It is present in exceedingly small quantities in the species in which it occurs, and is absent from some of the genera of the family. According to Church, it is absent in the species of *Schizorrhis*, in which the parts of the feathers which are in other genera coloured by turacin are here marked by white patches destitute of pigment. The great interest of the pigment is that it contains copper and not iron, but presents many interesting analogies to hæmoglobin. There is, however, no evidence that it can exist in both the oxidised and the reduced condition.   The pigment, like carmine, behaves as an acid, being readily soluble in dilute alkalies, but insoluble in acids.

Turacin is generally supposed to be confined to the plantain-eaters, but it has been also described by Krukenberg in one of the cuckoos (*Dasylophus superciliosus*).   The use, meaning, and origin are alike unknown ; its importance in coloration appears to be relatively slight, the feathers in which it occurs frequently showing bluish structural colour in addition to the red colour due to turacin.   It is somewhat interesting to note that the family of the plantain-eaters is an exceedingly small one of very limited distribution—it occurs in Africa only.

If turacin be boiled for a long time in air, it loses its red colour and becomes green, the change, according to Krukenberg, indicating the conversion into a new pigment.   This new pigment he describes as being devoid of copper, but containing a considerable amount of iron ; its spectrum shows a single band, instead of the two of turacin itself.   This green pigment was found by Krukenberg in the green feathers of *Corythæola cristata*, one of the Musophagidæ in which turacin is absent, and of *Corythaix albicristata*, one in which it is present.   This seems therefore to be one of those interesting cases of chemical relations existing between the different pigments of allied genera—a subject of which we know only too little.   Church is, however, inclined to doubt the existence of an independent green pigment.

Another interesting pigment of similarly restricted distribution is the red colouring-matter to which the red feathers of the male king paradise bird (*Cicinnurus regius*) owe their brilliancy. This pigment, called zoorubin by Krukenberg, occurs freely in various species of the birds of paradise, chiefly in the males, and has also been found in one of the Indian trogons (*Pyrotrogon diardi*) in the male, in the great bustard (*Otis tarda*), and in certain varieties of the common fowl.

Zoorubin is soluble only in dilute caustic soda, from which it is precipitated by the addition of acid as a dull brownish mass. Its solutions show no bands, but give two well-marked reactions. If cold concentrated sulphuric acid be poured cautiously into a test tube containing the solution, a blue or green ring forms at the junction of the liquids. Again, if the solution be rendered feebly acid and a trace of copper sulphate added, a bright cherry-red colour is produced. The pigment does not appear to contain iron or copper.

This list almost exhausts the known pigments of birds, and its two most striking features are, on the one hand, its uniformity, and on the other, the occurrence of peculiar and rare pigments like turacin in exceptional cases. It may, of course, be suggested that the impression of general uniformity is due to ignorance, and that many families of birds may contain peculiar and as yet undescribed pigments. There is, of course, no proof that this is not so, but at the same time the observations which have been made by Krukenberg and others tend to prove at least the very wide distribution of lipochromes and melanins, while they have failed to disclose any pigments of the uric acid group. The presence of

melanin pigment is perhaps explicable enough in
view of the great prevalence of these pigments in
Vertebrates, but what are we to say of the lipochromes? Is their presence in the feathers in some
families and apparent absence in others a sign of
the greater primitiveness of the first or not? If
their presence in feathers is associated with the
amount of oil in these structures, why are they
absent from the hair of mammals, which is also very
oily? We have also to consider the curious fact
that, while the muscles of fishes may be coloured by
red lipochromes, those of birds are not so coloured,
and the fat of birds is apparently always (?) coloured
with yellow, and not with the red lipochromes. Are
the reds formed from the yellow during the process
of the development of the feathers? These and
many other similar questions are suggested by the
study of the pigments of birds, and some at least
might be answered by a careful study of the pigments even of the species of a genus. To say that
the coloration is in each case produced by natural
selection obviously helps us little, for it can hardly
be supposed that the insignificant colour patches produced by turacin can have been of such supreme
importance as to determine the development of a
new pigment, while similarly in the birds of paradise
a red colour is sometimes due to zoorubin and sometimes not.

PIGMENTS OF BIRDS' EGGS

The colours of the egg-shells in birds are, as is
well known, often beautiful and varied. Rare as

blue pigments usually are among animals, blue and green tints are exceedingly common among birds' eggs, while various shades of brown, red, and yellow also occur.   According to Professor Alfred Newton (*Dictionary of Birds*, article " Eggs "), there is some reason to believe that for a time the eggs increase in brilliancy of colouring with each season until a maximum is reached, after which the brilliancy may again begin to decline.

The pigments of the egg-shells of birds have been investigated by several authors.   The important points upon which all agree are first, that the colouring is due to definite pigments ; and second, that these are derived directly or indirectly from hæmoglobin—results of much theoretic importance.   The interesting point is not that derivatives of hæmoglobin should be used in coloration, but why, if vivid and beautiful colouring-matters do arise in this way, they should not be employed in the coloration of the feathers.   It seems also generally admitted that even the ingenuity of that highly esteemed person, the field naturalist, is unequal to the task of explaining the colours of all birds' eggs upon the hypothesis of usefulness, so that from the theoretical point of view these pigments are of quite special interest.

Of pigments colouring eggs, Mr. H. C. Sorby describes seven with the following names and properties :—

1. Oorhodeine — a red-brown pigment of very common occurrence and great permanence.

2. Oocyan
3. Banded oocyan } blue pigments probably closely related, of which the second only yields a banded spectrum.

4. Yellow ooxanthine—a bright yellow pigment giving rise when mixed with oocyan to the bright permanent green so familiar in the eggs of the emu.
5. Rufous ooxanthine—a reddish-yellow pigment perhaps peculiar to the eggs of the tinamou.
6. A substance giving a banded spectrum but otherwise little known.
7. Lichenoxanthine—a brick-red pigment, possibly due to the growth of minute fungi.

As to the nature of the pigments, Krukenberg regards the blue and green colours as due to modifications of the bile-pigment biliverdin, and the brown and red colouring-matters as closely allied to iron-free hæmatin (hæmatoporphyrin). A more recent observer, Wickmann, regards all the pigments as originating directly from hæmoglobin. According to him, the pigments originate from the blood which fills up the *corpus luteum*. This blood stagnates and undergoes retrogressive metamorphoses which result in the formation of the pigments. He compares the process to that occurring in mammals, where there is a formation of hæmatoidin crystals in the *corpus luteum;* the difference may, perhaps, be explained by the diminished outflow of blood in mammals consequent on the greatly reduced size of the ova. According to Wickmann, the pigments formed in this way within the ovary are shed into the oviduct, and mingled with the materials of the shell in its uterine portion. If his observations are correct, they perhaps help to explain the facts noticed by Professor Newton (*op. cit*), that when a bird lays only two eggs, it not infrequently happens that all the available pigment is deposited on one, while the other may be colourless. Professor Newton gives the Golden

Eagle as an example of this. Wickmann further explains the differences in the pigments of the eggs of different birds as the result of differences in the composition of the blood. It is well known that in mammals the blood varies in different species, as is shown by the differences in the shape of the crystals of hæmoglobin, the colour of the plasma, and so on ; similar differences may express themselves in birds as differences in the products of decomposition.

For some criticisms of and additions to these statements of Wickmann, reference may be made to papers by Taschenberg and Von Nathusius.

If the pigments of the shell are iron-free derivatives of hæmoglobin, then the question of the fate of the iron thus set free becomes interesting. Krukenberg is of opinion that it may be used to colour the feathers in some cases ; he speaks of finding a large amount of iron oxide in the feathers of the lämmergeier, the feathers losing their dark brown colour after the removal of the iron.

MARKINGS OF FEATHERS

We have already touched upon the interesting questions connected with the markings of birds' feathers, but a general survey of the colour phenomena of birds would be incomplete without some further reference to them. It is unfortunate that there is so little certainty on the subject.

First, as to the origin of markings, and the simplest form of marking. On this point there are many suggestions, unfortunately, however, in most cases only suggestions. Häcker, in an interesting paper on the subject, adopts the view that longitudinal striping is the most primitive condition, that this tends to develop into a spotted condition by the suppression of portions of the stripe, and that the fusion of spots gives rise to cross-striping. Kerschener, on the other hand, regards cross-striping as the primitive condition from which spots are derived.   In point of fact, the distinction is perhaps less important than it seems, for Häcker's conception of waves of pigmentation passing down the shaft might equally be regarded as resulting in longitudinal striping, or in a very primitive form of cross-barring.   Häcker's observations were made chiefly upon nestlings of thrushes and chats (Turdidæ and Saxicolinæ), and also upon certain of the Limicoline birds.   His researches lead him to regard the most primitive form of colouring as that seen in some of the downs of the Limicolæ, such as *Podiceps rubricollis*, where there is merely a little pigment collected at the tip of an otherwise colourless down.   This is his *primary* pigmentation.   Most downs, however, show, on the other hand, in addition to this terminal patch of pigment, a basal pigmented area due to the process of *secondary* pigmentation.   In this way is produced the characteristic appearance of the feathers of young thrushes, where there is a pigmented downy area, and then a clear colourless area defined by a terminal pigmented band.   Besides occurring in the young of the thrushes and their allies, this type of coloration is found in the adults in the simplest feathers, such as those of the cheeks, the chin, etc. The primary pigment may form a dark spot at the apex of the feather, giving the plumage a spotted   ·

appearance, or it may spread out to form an apical
band which then gives the plumage a cross-barred
appearance ; of the two the first is the more primitive.
Beginning with these simple types of pigmentation
common to the downs both of the Limicolæ and the
Turdidæ, Häcker seeks to prove that the coloration
of the adult thrushes can be derived from this primitive type by various processes, especially the increasing importance of the secondary pigmentation.
Thus if the secondary pigmentation increase greatly
in importance, it may encroach upon the colourless
median area and, uniting with the primary apical
pigment, produce a uniformly coloured feather.
Again, the primary pigmentation may disappear,
and the colourless median area form a border to
the secondarily coloured feather, and so on.   It is
unnecessary to carry the consideration of Häcker's
theories beyond this point.   There is apparently no
doubt that the spotted appearance of the plumage in
young thrushes is a primitive condition, and the
nature of the pigmentation of the feathers in them is
therefore of great interest, but when the attempt is
made to derive more complex forms of marking
from these simple ones, there is great difficulty and
uncertainty.   A point of some interest is the question
whether there is any relation between the structure
of special regions of feathers and the characteristic
pigmentation of these regions.   It is at least certain
that there is much constancy in the association of
special types of colour with special regions of the
feather.   The nature of the association we shall
consider in the next chapter in connection with the
colours of certain families of birds.

CHAPTER XIII

THE COLOURS OF BIRDS (*Continued*)

The Structure of Feathers—Relation between Structure and
Colour—The Colours of Sun-birds, Humming-birds, and
Birds of Paradise, their Distribution and Characters—
Markings of Kingfishers — General Characters of the
Colours of Birds—Meaning of Colour in Birds.

HAVING in the previous chapter considered some
general aspects of the colours of birds, we shall now
proceed to study the coloration of special families in
detail. In order to make the descriptions readily
comprehensible, it will be first necessary to briefly
revise the structure of feathers.

Feathers are outgrowths of the epidermis, formed,
like all such outgrowths, of the substance keratin.
They differ according to their function, and the part
of the body in which they occur. Thus there are
the quill-feathers which occur in wings and tail, the
general contour-feathers which cover the surface of
the body, and the downs or soft under feathers,
which are often abundant on the breast. All these
either contain pigment or are filled with bubbles of
air and so display a white colour. Before proceeding to describe the distribution of pigment in these

feathers, we shall consider their characteristics in detail.

We shall take first a quill, such as one of the primaries of the wing (see Fig. 1). Such a feather consists of a central axis or stem bearing on its upper portion a large number of lateral growths—the barbs. The lower naked part of the axis forms the quill, while the whole of the remainder of the feather is known as the vane and consists of a central rachis

FIG. 1.—Feathers of sun-birds to show relation between colour and shape. The quill-feather was uniformly coloured except for a slight edging of metallic colour at one side; the short feather shows three zones—a terminal metallic zone, a median dark-coloured and slightly V-shaped zone, and a downy basal zone.

and lateral barbs. The barbs of the vane are closely connected together, and on pulling them gently apart, it is possible to see that they bear on either side innumerable small processes, the barbules. The barbs cling together because the barbules are locked to one another in a manner presently to be described.

Each barb bears two rows of barbules, and one of these rows points to the tip of the feather, and the other to its outer or inner edge. The latter is called the proximal, and the former the distal row. Each barbule consists of a flattened process which appears to be twisted upon itself at about the middle of its length. Its proximal part has therefore the appearance of a flattened lamina and its distal of a filament, as owing to the twist the edge only is in the plane of the lamina. Now in the barbules of the distal series, the filamentous region bears a series of hooklets and slender processes which fit into a groove and notches developed in the lamina of the proximal barbules. Each set of distal barbules is thus hooked into a set of proximal barbules, so that each barb is locked to its neighbour. When the uniform surface of the feather vane is destroyed by forcibly separating the barbs, the hooklets are pulled out of the groove in which they lie. When the feather is restored to its original condition by smoothing with the fingers, the hooklets are slipped back into their original position.

The barbs just described constitute the greater part of the vane of a quill-feather, but at the base of the vane there will usually be found a number of barbs of very different appearance. These are the downy barbs, and they are characterised by the fact that they are quite unconnected, and that their barbules are usually very long and slender, so as to be far more conspicuous than the barbules of the vane proper. These barbules bear no hooklets, the twisting is less obvious, and the appearance of length is given by the great development of the filamentous region. Barbs bearing barbules of this type are the only ones present in down feathers; these are further characterised by the shortness of their axis, and are rarely important in coloration, except in the young.

The small feathers which cover the surface of the body differ in several respects from quills. They are much shorter, the quill region is practically absent, the rachis is reduced in length and thickness, and the downy region tends to be more fully developed. The result of this shortening of the axis is that the barbs tend to radiate from a common point, while in feathers with elongated axis they are, roughly speaking, parallel to one another. This has an important bearing upon the coloration, for it is obvious that if the barbules standing near the ends of the barbs tend to exhibit special colours, then the colour will form a transverse band on short feathers, a longitudinal band on long feathers. Similarly the median barbules will form a transverse band on short feathers, a V-shaped marking on long feathers; both of these actually occur (see Fig. 1).

RELATION BETWEEN STRUCTURE AND COLOUR

We have already seen that there are three great sets of colour phenomena displayed by the feathers of birds :—(1) The feathers may show beautiful and complex markings in brown, gray, and black; (2) they may display vivid optical colours; or finally (3) may contain brightly coloured pigments, usually of the nature of liprochromes. Of these three, the first occur equally in long quill-feathers and in the short contour-feathers; they have no obvious relation to the structure of the coloured parts, and as already seen, little is known of their meaning or course of evolution.

(2) The optical or structural colours are divided into subjective and objective. The changing subjective colours occur only in the barbules, and require the presence of a large amount of dark pigment for their full manifestation. Objective colours like green and blue are confined to the barbs and do not occur in the barbules, and (3) the bright pigments occupy the same position. In general terms, therefore, we may say the barbules always contain a certain amount of dark pigment, and when this is in excess and the structure is modified metallic colours arise. The barbs, on the other hand, may contain dark pigment, may show objective optical colours, or may contain bright pigments. The variations which produce these colour phenomena are much commoner in the general feathers of the body than in quills; they do not usually occur simultaneously, and the appearance of any one set of colours is associated with an increased development of the special region of the feather with which the colour is associated, as of the barbs, a portion of the barbules, and so on. It may be that this is in part the explanation of the fact that, apart from the development of markings, quill-feathers are slow to vary in colour, and are rarely brilliant. Colour brilliancy is associated with a special development of some individual region of the feather, and it is essential for the purposes of flight that there should be a harmonious development of all the parts

of the quills, and no specialisation of particular areas. Therefore any tendency to the development of brilliant colouring in the wing-quills would be checked by the resulting injury to flight, and so to the well-being of the species.

It will be noted that in the following descriptions red and yellow colours are always ascribed to the presence of lipochrome. We have already seen that, according to Gadow, yellow may at times be an optical colour ; in the cases discussed, however, the presence of yellow lipochrome has either been directly proved, or is assumed from the simultaneous occurrence of a red colour, which is always due to lipochrome pigment. The thesis here put forward as to the relation existing between brilliant colouring and variation in feather structure, we propose to develop by a consideration of the colour phenomena in sun-birds, humming-birds, and birds of Paradise.

COLOURS OF SUN-BIRDS

The Nectariniidæ or sun-birds are a family of mostly small birds often with brilliant colours, inhabiting Africa, and India, and the Malay, where they seem to replace the American humming-birds. Their beauty and their habit of frequenting flowers have caused them to be frequently confounded with true humming-birds, but they are not in any way related to the latter. The bright colours are almost entirely confined to the males, and are by them acquired with extreme slowness, so that birds are said to be not infrequently seen mated while the male is still in a sort of hybrid plumage. Birds in this condition are peculiarly ugly, as the bright metallic feathers occur scattered among the dull youthful plumage. If the colours of the male exercise as important an influence on the choice of the female as is commonly asserted for birds, the female sun-bird must also be assumed to be possessed of much faith and foresight. Some at least of the ornamental feathers of the male are cast almost as soon as the breeding season is over.

Nature of Bright Colours

The bright colours of the sun-birds are due either to lipochrome pigments or to metallic structural colours, belonging to Gadow's group of subjective structural colours.

1. *Pigmental Colours.*—The colours due to lipochrome pigments are either yellow or bright scarlet-red. Brilliant patches of red or yellow feathers frequently occur on the throat or on the ventral surface, or dorsally at the root of the tail. The feathers so coloured are always short contour-feathers and not quills. The bright colour is confined to the apical part of the feather, the base being grayish or white, and the pigment occurs as usual only in the barbs. The barbules, if present, are grayish, but most frequently they are rudimentary or absent, so that the visible part of the feather consists of the diverging naked barbs, containing a considerable amount of bright pigment.

2. *Structural Colours.*—The metallic colours of the sun-birds occur on feathers arranged in special patches on the head and throat, or as transverse

bands near the tip of the general contour-feathers, or as longitudinal bands at the edges of the quill-feathers.

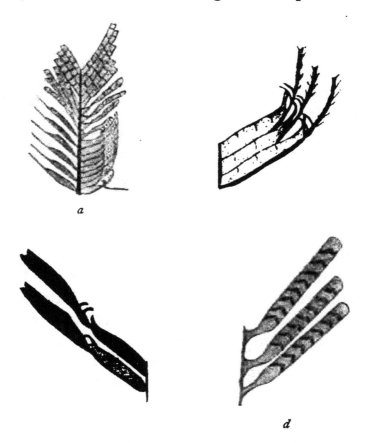

FIG. 2.—Barb and barbules from feathers of sun-birds, magnified to show the peculiar structure of metallic barbules. *a* is a barb bearing both proximal and distal barbules, the lower barbules being partially, the upper completely metallic. *b*, non-metallic distal barbules from a tail-quill of a sun-bird, showing lamina, hooklets, and filamentous region. *c*, partially metallic distal barbules ; the colour is confined to the enlarged filamentous region, but the lamina and hooklets persist unaltered. *d*, completely metallic barbules, with no trace of hooklets or lamina.

In tint they are usually green, blue, violet, or reddish-violet, yellow or red structural colours being absent.

The colours are produced by a modification of the barbules of the metallic feathers. We have already described the general structure of barbules and noticed that each is divided into two regions—a proximal flattened region which may be called the lamina, and a distal slender region which, from its appearance, may be called the filamentous region. Now as in sun-birds the metallic colours are usually confined in quill-feathers to a lateral stripe, it is obvious that it is possible to obtain a single barb which bears both metallic and non-metallic barbules. If we examine microscopically a non-metallic barbule, we shall find that it exhibits the ordinary structure of a barbule, and shows quite distinctly the division into two regions separated by a twist (see Fig. 2, *b*). The metallic barbules (*d*), on the other hand, are of quite different appearance, being broad, flattened, club-shaped bodies supported on a short stalk, and containing abundant dark pigment. Close examination of the barb (*a*) shows that the metallic and non-metallic barbules are not perfectly sharply defined, but tend to pass into one another. Thus, as we follow the non-metallic barbules upwards, we find that the lamina diminishes in size, while the filamentous region becomes flattened, broader, and larger, at the same time losing its slender processes (*c*). Finally, the lamina becomes so much reduced as to form only the short stalk of the metallic barbules, while the distal region becomes modified into the club-shaped body, and is then completely devoid of hooklets or processes (cilia). These club-shaped barbules further exhibit a series of cross bars which, according to Gadow, are a series of compartments overlapping like the tiles of a roof. The ultimate causation of the physical colour Gadow ascribes to the transparent sheaths of keratin covering these compartments, which he thinks act like a series of prisms. An important point in connection with these metallic barbules is, that they are so modified that both hooklets and folds are completely lost, and therefore there is no connection between the barbules or the barbs. Metallic feathers of this type have therefore a peculiar looseness of texture which is, for example, very obvious in the ornamental feathers of the peacock; the solidity of the flattened metallic barbules gives, however, to such feathers an appearance quite different from that of ordinary downy feathers, in which also the barbs are unconnected. The unconnected nature of the barbs is of especial interest, because it would render the feathers quite unfitted for purposes of flight if the variation were to occur in quill-feathers. In sun-birds it is usually the contour-feathers which are metallic, rarely the tail-quills, and apparently never the wing-quills.

Development of Colours.—The types of coloration already described in the sun-birds are seen in the specialised feathers, especially of the male. In the unspecialised feathers, such as the general contour-feathers of the female, we find what may be regarded as the primitive condition. These feathers are of a dull olive colour, and are divided into three regions—a basal downy region usually of an ashy colour, a median slightly V-shaped region in which the barbules have a very close texture and are of a brown colour, an apical region in which the barbules are unconnected, slightly modified, and faintly pigmented with dark pigment, while the barbs stand out as being of a dull yellow colour, apparently produced by a mixture of lipochrome and melanin. This type is seen in most of the females and in the males of the inconspicuously coloured species. Taking such a feather as a starting-point, we may have divergence in two directions. In the first case, the lipochrome pigment may increase greatly in amount, and colour the barbs very deeply, while the dull barbules do not become brightly coloured, but tend to become rudimentary and disappear; thus we get the bright red or yellow patches formed. On the other hand, the lipochrome may get swamped by the development of a large amount of melanin, which occurs not only in the barbs, but also in the barbules. At the same time, the modified barbules near the ends of the barbs progress further in the direction in which they have begun to develop, become converted into completely metallic barbules, and thus give rise to the band of metallic colour seen at the ends of the general contour - feathers of many males, e.g. *Nectarinia famosa.* When, as in *Cinnyris frenatus*, the ventral surface of the male has both metallic colour and bright pigmental colour, it is possible to find individual feathers displaying both tendencies—that is, with naked yellow barbs at the tip, then metallic barbules placed on a dull-coloured region of the barb, and then the covered unspecialised region of the feather.

In quill - feathers the tendencies as to colour evolution seem slightly different. The quills of the females, of the unspecialised species, and of the wings of perhaps all species are a dull brownish colour, with a tendency to exhibit a longitudinal edging of olive colour in which the barbules are pale-coloured, unconnected, and slightly modified. This pale band is sometimes replaced in the greater wing-coverts of the male by a dark brown band ; in the tail-quills of the male usually by a metallic band which, in the case of the central rectrices, may invade the whole vane. The development of lipochrome colour or of transverse bands of colour does not occur in the case of quill-feathers. The latter is due to the fact that in sun-birds it is only the barbules which stand near the distal end of the barb which tend to become metallic, the result being the formation of transverse stripes of bright colouring on short feathers, and longitudinal stripes on long feathers, the type developed having a definite relation to the length of the feather (see Fig. 1). This peculiarity is apparently the result of the fact that in metallic barbules the lamina tends to disappear, and this seems to occur typically only in downy barbules or in the barbules standing near the apices of the barbs. Downy barbules never become metallic, so that it i_:only the apical barbules which can become metallic, and give rise to a band of colour.

From the above description it is obvious that the development of brilliant colouring in sun-birds is certainly associated with modifications of feather structure which cause the feathers to deviate more or less completely from the primitive type of feather structure.

Colours of Humming-birds

Turning now to humming-birds, we find that here pigmental colours are of relatively little importance, while structural colours attain an extraordinary beauty and brilliancy. Further, we find that the place of the pigmental colours of sun-birds in contrasting with and showing up the metallic colours is taken in humming-birds by black and white. White especially is often of great importance in producing the general effect of beauty.

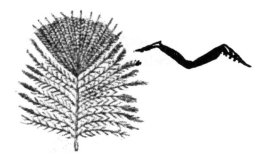

FIG. 3.—Metallic feather from the throat of a humming-bird. The colour is confined to the central pigmented patch and is there exceedingly bright. The barbs are continued beyond the metallic region, but are then without barbules. The figure at the side is a section through the barb to show the method of insertion of the barbules which produces strongly marked ridges on the surface, so that the barbs lie at the bottom of a trough.

In humming-birds metallic tints occur in both sexes, but are usually more brilliant in the male. They very frequently occur on the general contour-feathers, the colour being then often a bronze-green, which is not sharply confined to a transverse band, but fades away gradually behind. The metallic

colours which are especially characteristic of humming-birds, however, occur, as is well known, in
patches of extraordinary brilliancy either on the
head as a crest, or on the lower surface, especially of
the throat. The feathers forming these patches are
peculiarly modified, and may display any of the
colours of the spectrum including ruby-red and
golden-yellow—the colours which are so markedly
absent from the metallic feathers of sun-birds. The
rectrices of humming-birds not infrequently display
metallic colour, which may be distributed over the
whole feather, or may be limited to a transverse band
near the tip. Longitudinal bands of metallic colour
like those of the sun-birds do not seem to occur.

Pigmental colours among humming-birds are not
remarkable for brightness of tint, being usually shades
of gray or dull brown. The only marked exception
is the colour called by systematists "rich chestnut"
or "cinnamon," which is often limited to the males,
as, for example, in *Eustephanus fernandensis*. In
this connection it may be noticed that not only
are metallic tints almost invariably absent from the
wings, but where, as in the above species, the male
as compared with the female is characterised by the
development of a special pigmental colour, this pigment is entirely absent from the wing-quills, though
present in the wing-coverts.

As an exception to the general rule that the
humming-birds display great brilliancy, we find that
the so-called "hermit" forms which live in the deep
shades of the forests are only soberly tinted, with
little metallic colour ; of these the genus *Phæthornis*
may be taken as a type.

Structure of Metallic Feathers. — The brilliant metallic feathers of the head region of many male humming-birds are in several respects very peculiarly modified. They are very short, much rounded, and overlap one another ; the surface is strongly metallic and marked with deep ridges (see Fig. 3). A further point of interest is that the barbs, quite devoid of barbules, are prolonged as a delicate fringe beyond the apex of the feather. While for further details I must refer to my paper on the subject, we may simply notice that the metallic colouring is here not produced by a modification of the distal portion of the barbules, but by a deeper pigmentation and a structural change in the proximal region. The result of this is that the metallic colour in humming-birds tends to appear first in the middle region of the body feathers—that is, the region where the barbules tend to attain their maximum development, and not at the tip of the feathers as in sun-birds. This primitive condition is well seen in the breast feathers of the female of *Eustephanus fernandensis*, and of both male and female of *E. galeritus*. Here we have white or dull-coloured feathers, with a central spot deeply pigmented, and displaying a varying amount of metallic colour. The increasing specialisation of the metallic region is accompanied by a gradual retrogression of the apical region which is eventually represented only by the slender naked barbs.

The metallic modification of the feathers in humming-birds is therefore not accompanied by any change which affects the locking together of the barbules, and so the adaptability for purposes of flight ; it differs in this respect sharply from the modification seen in the sun-birds where the metallic barbules are entirely unconnected. We can thus understand how it is that the quills of humming-birds may display structural colour without their efficiency being in any way impaired. The fact is also readily explicable on structural grounds, if we · recollect that it is the mid region of the feather which tends to become metallic, and it is this region which is most fully developed in quills. In humming-

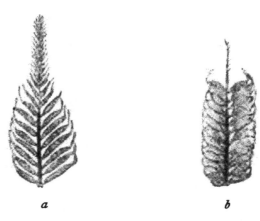

a   *b*

FIG. 4.—Metallic barbs from feathers of humming-birds, magnified. *a* is from a feather belonging to a female of *Eustephanus fernandensis*, the barbules at the base are metallic, but the barb also bears rudiments of barbules at its tip. *b*, a barb from the feather shown in Fig. 3 ; all the barbules are metallic and the tip of the barb is naked.

birds the colour differences between the male and female, or between specialised and unspecialised species, are thus largely the result of an increased amount of melanin pigment in the brilliant forms, accompanied by a process of structural modification.

The type of metallic colour seen in the humming-birds is of much interest, because it has not been described outside of the family. In other cases we find that metallic colours in birds are of the type described in sun-birds, *i.e.* are due to the conversion of the distal portion of the terminal barbules into a club-shaped body consisting of a series of overlapping compartments, the process being accompanied by the total suppression of cilia and hooklets. Such a modification of structure is apparently of very common occurrence in birds, but does not give rise to metallic colour unless there is a simultaneous development of a large amount of black pigment.

COLOURS OF THE BIRDS OF PARADISE

We may now for a little pass to the consideration of the birds of Paradise, which on account of their greater size afford more obvious illustrations of some colour problems. The birds of Paradise as a group exhibit an extraordinarily specialised type of coloration, the specialisation being visible alike in colour and in structure. Though probably nearly allied to the crows, the development of melanin pigment is here less remarkable than the display of bright pigmental colours. These are in part due to lipochromes, but in part, as we have already seen, to the pigment zoorubin, which is almost confined to the group. Further, we have not only the display of tufts and crests of additional feathers, as in humming-birds, but we find that these feathers are modified in every conceivable way, sometimes being reduced to mere wires, and at other times displaying brilliant metallic colours. Among many of the birds of Paradise the metallic colours are of somewhat limited distribution, contrasting with the pigmental colours rather than

forming the basis of coloration.   In the nearly related
rifle-birds, on the other hand, the pigmental colours
have disappeared, and the great predominance of the
melanins is associated with the development of the
most gorgeous metallic colour, set off by the velvety
blackness of other feathers ; a similar type of coloration occurs in the genera *Parotia* and *Astrapia* among
the true birds of Paradise ; in these the speckled
plumage of the female is very noticeable.   As usual
throughout the beauty of colouring is largely confined
to the adult males, the females and young males
being relatively unadorned.

The birds of Paradise, as is well known, inhabit
the Malay Archipelago, and a full description of the
family, including an account of the native methods
of obtaining them, will be found in Mr. Wallace's
account of his travels in that region.

The great bird of Paradise, called *Paradisea
apoda* by Linnæus, who described it from a specimen
preserved after the native method, and therefore
without its feet, may be chosen as an example of
one of the prevalent types of coloration.   In this
bird the quill-feathers of tail and wing, and the
feathers which cover the greater part of the back, are
a dull brown colour, showing little specialisation of
colour.   In the feathers of the back the barbs are
devoid of barbules near their apices, but show no
other specialisation.   At the sides of the body are
the beautiful erectile tufts of feathers, which give the
bird half its beauty.   These consist of long drooping
feathers, pinkish-white in colour, with a tuft of short
bright yellow feathers at their base.   The elongated
feathers have much elongated barbules, a structure some-

what resembling that seen in downy feathers, and like many downy feathers they have very little pigment. Among these long feathers are the so-called " wires," which are feathers with all the parts save the rachis suppressed. The yellow feathers have no barbules, and the barbs are smooth, dilated, and brightly coloured with lipochrome pigment. The same pigment and the same feather structure is found in the bright yellow feathers forming the crest. Round the base of the beak and extending over the throat there is a band of green metallic colour produced by very much shortened feathers, in which the barbules have undergone the same modification as those of the metallic feathers of sun-birds. Speaking generally, we may say that this bird shows more tendency to develop additional plumes than any great brilliancy of colour, but when bright colours are developed, their development is associated with a tendency to suppression of the barbules, and to dilatation of the naked barbs.

The king Paradise bird (*Cicinnurus regius*), on the other hand, shows less tendency to develop additional tufts but much greater brilliancy of tint. The female is a dull grayish tint, with a speckled breast, the male is brilliant red on the back, with a metallic green band separating the red head from the pure white of the ventral surface. The tail contains two much elongated wires displaying a brilliant green colour at their curled tips. The red feathers are coloured by the peculiar pigment zoorubin, which is practically absent from the female. The feathers containing it have as usual naked barbs, which are smooth, dilated, and polished, so that as compared with the general contour-feathers of the preceding species they are modified both as to structure and pigmentation. The quill-feathers are mostly dull in colour, but a close examination shows that this relative dulness is due to the fact that, while the barbs are as before coloured with bright red zoorubin, the barbules contain a dull brown pigment, the result being to greatly diminish the brilliancy of colour. Certain of the quills have bright scarlet longitudinal bands at their edges ; this is due to the fact that here the barbules are absent and the bright red barbs have their full effect.

It would be tedious to go on to discuss in detail the coloration of the rifle-birds, but we may briefly notice that here, associated with the development of a large amount of melanin and the loss of the lipo-chromes, we have also the loss of the tendency to suppression of the barbules ; here these tend to become greatly specialised, and to develop metallic colour. In the elongated metallic feathers of the throat of e.g. *Ptiloris magnifica*, we have further the development of those V-shaped bands of which we have already spoken.

The examples given above have been taken from birds exhibiting bright pigmental colours or subjective structural colours ; perhaps we may be allowed further to give some illustrations with regard to the objective structural colours, like blue and green. A blue colour is always entirely confined to the barbs of feathers, and is often associated with a suppression of the barbules ; it only appears on exposed parts of feathers.

MARKINGS OF KINGFISHERS

Blue and green structural colours are admirably displayed in the family of the kingfishers, which show also a gradual progression of colour. Thus in *Ceryle rudis* the feathers are dark brown or black, more or less irregularly spotted with white, but with the white showing a distinct tendency to form a transverse band at the tip of the feather. In *Ceryle guttata* the feathers are regularly cross-barred with dull blackish-brown and white. In *Carcinentes melanops* in the wing-coverts the covered part of the feather is striped black and white, but the terminal bar of white is replaced by blue. In the tail-quills the under surface is usually black and white and the upper surface blue and black. Where there is partial overlapping of the quills, one side of the vane may be black and white and the other exposed side black and blue. The blue patches occur in positions corresponding to the white ones, but are larger and show a tendency to fuse together. In the case of quill-feathers, the blue is confined to the barbs but the barbules are still present, and their dull colour somewhat diminishes the brilliancy of the blue. On the general contour-feathers, on the other hand, the development of the blue colour is associated with a suppression of the barbules, while the barbs as usual tend to become dilated and polished.

While in many kingfishers blue and black are the dominant colours, in some the blue is replaced by green. Thus in *Halcyon lindsayi* a yellow colour is common on various parts of the feathers, and where structural colour occurs, it is here green and not blue. Many of the feathers of the back are black, cross-barred with yellow, and here the terminal cross-bar is wholly or in part replaced by green.

GENERAL CHARACTERS OF COLOURS OF BIRDS

These illustrations of colour phenomena in birds, if they do not explain the development of bright colour, may perhaps at least shed some light on the problem. They show that the development of brilliant colour and structural modification go hand in hand ; that brilliant pigmental colours tend to be confined to the barbs and are often associated with the suppression of the barbules ; that melanin pigments may be present in large amount in both barbs and barbules ; and that their presence in the latter is often associated with a structural modification which gives rise to optical colours ; that the closeness of the association between the deposition of pigment in any region of a feather and the special development of that region is such as to prevent in the general case the feathers of flight acquiring great brilliancy of colour. Facts of this kind surely tend to prove the definiteness of variation ; they should, at least, be allowed for by those who discuss the questions connected with the origin of colour.

MEANING OF COLOUR IN BIRDS

As to the meaning of the various types of colour in the physiology of birds we can say very little.

Those associated with the presence of melanin pigment are here as elsewhere less inexplicable than those due to lipochromes. They are certainly far more conspicuous in the males than in the females, and may therefore be ascribed more or less directly to the greater vitality of the male, which expresses itself in more rapid metabolism and increased production of pigment. But with the lipochromes the case is different ; their constant association with fats makes it difficult to regard them as products of destructive metabolism. Again, though in many cases they are only conspicuous in coloration in the males, yet the peculiar case of the green and red parrots (p. 252) seems to show that in some instances they may be more abundant in the females. It may, of course, be suggested that the lipochromes in the female are largely used up in the colouring of the yolk, while in the male they may colour the plumage ; but it is still difficult to account for the virtual absence of lipochrome from many families, as the crows, the humming-birds, and so on—families certainly highly specialised in other respects. Nor can we regard the presence of lipochrome as indicating want of specialisation in view of the fact that, as in the birds of Paradise, they may be absent or present within the limits of a family, without obvious differences in the amount of specialisation. In view of the absence of lipochrome from the cuticular structures of mammals the question is of much interest. The absence of derivatives of waste products in the cuticular structures of birds as compared with insects is, of course, probably to be associated with the well-developed, excretory, and vascular systems in a bird as compared with those of an insect. The great differentiation has now rendered it impossible for nitrogenous waste products to be directly employed in coloration.

CHAPTER XIV

THE COLOURS OF MAMMALS AND THE ORIGIN OF PIGMENTS

Coloration of Mammals—Pigments of Mammals—Colour of the Hair and Skin in Man, and its Bearing on General Problems—Origin of Melanin—Pigment and Waste Products—Experimental Evidence—Conclusions—Criticism of these Conclusions.

WE have already remarked on the familiar fact that mammals are rarely remarkable for brilliant pigments, the prevailing colouring-matters being the dull-coloured melanins. The statement is, of course, true only of the colours of the skin, for bright pigments do occur in the tissues; thus hæmoglobin colours not only the blood but most of the muscles, and yellow lipochromes occur often in quantity in the fat, in the plasma of the blood, in the muscles, and so on. Of these the hæmoglobin of the blood is an important factor in coloration in the white races of mankind, and when associated with certain peculiarities of the structure of the epidermis, gives rise to the bright tints of the callosities of many monkeys, of the face of the mandrill (*Cynocephalus maimon*), and so on. Under normal conditions the decomposition products of hæmoglobin, unless melanin be one of these, are of no importance in producing external colour ; nor do the lipochromes ever appear to occur in the epidermis or cuticle. Optical colours except white are rare in mammals, but true metallic colours occur in the Cape golden mole (*Chrysochloris*). In this little insectivore the fur especially of the upper surface displays " a brilliant metallic lustre, varying from golden-bronze to green and violet of different shades " (Flower). The exact causation of the colour appears to be unknown.

In mammals generally the beauty of the colouring is dependent upon the unequal distribution of the melanin pigments, which are very frequently so arranged as to produce the effect of stripes or spots. There are several papers upon the origin, relations, and meanings of these markings, but all are too purely theoretical to demand detailed notice here. An account of them will be found in the works of Wallace, Eimer, Bonavia, and others.

Among the general colour characteristics of mammals, we should notice the tendency of certain variations to recur constantly in many different orders ; such are the deepening of the tint (melanism), the disappearance of the pigment (albinism), the prevalence of a sandy colour in mammals inhabiting deserts, and so on. Melanic varieties are seen not infrequently in the leopard (*Felis pardus*), especially in Southern Asia ; they seem to occur quite sporadically. A very interesting point about these black leopards is that, in certain lights, the markings characteristic of the leopard can be seen on the black ground like the pattern on " watered silk." This shows that these markings are not wholly determined by the amount of pigment present in the hairs, there must be also some additional cause.

Albino varieties occur occasionally as a sport, especially under domestication, but many mammals are naturally white. As is well known, certain Arctic animals, *e.g.* the polar bear, are always white, others only turn white in winter, *e.g.* the Arctic fox. The change of colour in these cases is associated with the development of numerous air-bubbles in the hair. It would seem that in some cases this is not accompanied by a destruction of the pigment, which is merely concealed by the air-bubbles.

For further particulars as to the characters of the colours in mammals, reference should be made to the text-books, and for the markings to Eimer's papers.

The pigments of mammals have been relatively little investigated, but there is probably great uniformity throughout the group. Leydig describes uric acid compounds as occurring in the skin of *Chrysochloris*, and regards them as factors in the coloration, but in general the colours are apparently due only to the melanins.

Colour of the Hair and Skin in Man

In connection with the pigments, a few remarks upon the colour of the skin and hair in our own species may not be out of place, especially as the questions connected with it have considerable bearing upon general problems. As is well known, the

dark races of mankind owe the colour of their skin
to a black pigment deposited in the deeper layers of
the epidermis—a pigment which is practically absent
in white-skinned people. The varying tints of the
hair are also due to the varying amounts of the same
dark pigment deposited in it. That differences in
skin-colour often correspond to profound racial differences is familiar in a rough way to every one, but
there are some interesting facts which tend to show
that even apparently slight differences in intensity of
pigmentation may correspond to relatively vast constitutional differences.

We propose to confine our study of the question
to variations in the colour of the hair and eyes in
the white-skinned peoples, where the data are most
easily obtainable. The first point of interest is the
curious fact shown by Galton's statistical researches
that among ourselves there is little tendency for the
dark and fair strains to mingle, "to be swamped by
intercrossing," in the current phrase. The children
of parents of whom one is dark and the other fair
will as a rule either have dark *or* light eyes, only
rarely will they have eyes of medium colour (*Natural
Inheritance*, p. 139). In Mr. Bateson's words, the
variations are discontinuous.

The next point is that, according to Dr. Beddoe's
prolonged observations, "the colour of the hair is so
nearly permanent in races of men as to be fairly trust-
worthy evidence in the matter of ethnical descent, and
nearly as much may be said for the colour of the iris"
(*The Races of Britain*, p. 269). His observations
further show that the dark-haired people correspond
roughly to the Gaelic and Iberian stocks, while the fair-

haired belong to the Teutonic races ; in other words, the difference in hair-colour corresponds to all those profound mental and moral differences which separate Celt from Saxon. That the mental and moral differences are associated with physical ones hardly needs proof to the biologist, but there are fortunately some exact observations. During the course of an extended series of observations on the specific gravity of the blood, Dr. E. Lloyd Jones found that this was markedly greater in dark-eyed persons than in light-eyed ones, and he is of opinion that the difference is fundamentally a racial one. Further, there is reason to believe that the dark-haired people are better able to stand prolonged dosing with drugs like mercury than the fair-haired ones ; and, according to Beddoe, the dark-haired persons in Britain are more prone to phthisis than the fair. It thus seems that just as the phthisical tendency and the other characters tend to eliminate the dark people from cold climates, so apparently the fair people are less fitted to survive, or at least less likely to become dominant, in hot countries. Facts of this kind have probably an important bearing upon the coloration of mammals in general. The constancy of the coloration, and the closeness of its connection with the constitution, are at least of much interest in relation to the general question.

THE ORIGIN OF MELANIN

As to the direct relation of the amount of pigment to the general metabolism, many would say that the pigment is directly derived from the hæmoglobin of the blood, and that, therefore, its amount is a direct measure of the rapidity of the degenerative changes occurring in the hæmoglobin. From some recent work it would, however, appear that there is not this direct relation between the pigment and hæmoglobin. Drs. John Abel and Walter Davis, in the course of a laborious investigation on the pigments of the negro's skin and hair, found that the pigment granules of the epidermal cells contained a substratum of non-pigmentary substance, apparently of the nature of a highly resistant proteid. When this proteid is removed the pigment is readily soluble in dilute alkalies, from which it may be precipitated by acids. It contains carbon, nitrogen, oxygen, hydrogen, and sulphur, but in the pure state very little iron. It is the presence of sulphur without any considerable amount of iron which, in the opinion of the authors, makes an origin from hæmoglobin very doubtful. The proteid which is also present in the pigment granules contains a considerable amount of iron as well as of other inorganic constituents. Floyd showed in 1876 (*Chem. News*, vol. xxxiv. p. 179) that the skin of the negro contains about twice as much iron as the white skin, but this is apparently due to the proteid and not to the actual pigment itself. The investigators are of opinion that the pigment originates from some proteid of the blood or "parenchymatous juices." Similarly Dr. Sheridan Delépine considers that melanin is elaborated out of the plasma of the blood and is not a derivative of hæmoglobin. On the other hand, he is of opinion that hæmoglobin itself is perhaps manufactured from some "antecedent, variety, or derivative of melanin," a somewhat interesting conclusion.

PIGMENT AND WASTE PRODUCTS

The question as to whether the pigments of mammals are to be regarded as products of destructive metabolism is one which has considerable bearing on the general question. In considering particular cases of pigmentation we have again and again come across suggestions to the effect that the pigments of organisms are effete substances incapable of serving directly useful purposes, which may be stored up in the cutaneous tissues, and so give rise to coloration. In considering these suggestions in a little more detail, we may, in the first place, provisionally exclude cases like that of the Lepidoptera, where the pigments, in some cases at least, are definitely excretory substances. Our immediate concern is not with these, but with the numerous kinds of pigment which are different from the ordinary waste products of the organism in which they occur, and which have not been proved to have a genetic connection with these. Such pigments have not infrequently been described as waste products, and it is necessary for us to consider how far this is justifiable.

EXPERIMENTAL EVIDENCE

In the first place, it is interesting to note that the suggestions have been usually made in connection with physiological experiments on leucocytes. The modern doctrine of the physiological and pathological importance of leucocytes and phagocytes, with which the name of Metschnikoff is so honourably associated, has been largely founded on results obtained from the injection of foreign substances in suspension or solution into the body. The injected substances are usually colouring - matters for convenience of observation, and the result has been to prove that they are systematically removed by leucocytes from the general cells of the body ; and either eliminated through the excretory organs or stored up in various parts of the body, where they may give rise to artificial coloration. Now we have already frequently seen that natural pigmentation may result from the emigration of pigmented connective tissue cells from the deeper tissues outwards to the skin. This occurs, for example, in the leech, and, according to Kölliker, is true for all Vertebrates. This being so, it is eminently natural that the physiologists should draw a parallel between these natural pigmented "wandering cells" and the pigmented leucocytes found after injection of colouring-matter, and regard the former as active agents in eliminating the normal waste products. The necessity for finding a physiological justification for the continued production of pigment being so obvious, the suggestion once made has been eagerly adopted by many.

The simplest case is that in which the introduced pigment is injected into the alimentary canal, and its subsequent fate compared with that of the pigments normally occurring in the cells connected with this. The following table shows the results obtained from some of these experiments :—

PIGMENTS INTRODUCED INTO ALIMENTARY CANAL

Organisms.	Situations in which introduced Pigments are found.	Natural Analogues.	Observer.
Polyzoa.	In "hepatic cells" of gut.	Brown pigment of these cells.	Harmer.
Crustacea.	In cells of hepatic cæca, ultimately in fæces.	Pigment normally found in hepatic cells and fæces.	Cuénot.
Capitellidæ.	In cells of gut, ultimately in skin.	Pigment normally occurring in cells of gut and in skin.	Eisig.
Oligochæta (*Tubifex*).	In the so-called chloragogenous cells covering the intestine, ultimately in the skin.	(?) Pigment of chloragogenous cells and of skin.	Cuénot.

In these cases the Capitellidæ are the only forms in which the introduced pigments come to have a marked effect on the coloration. Another case in which the introduced pigment is important in this way is perhaps the case of caterpillars, where the pigment introduced with the food reaches the connective tissues and so the skin. In the general case, however, it would seem that the pigments which normally result from the activity of the liver or "hepatic cells," as well as pigments artificially introduced into the gut, are usually directly eliminated by means of it, and do not become important in coloration. It may be that worms are an exception to this rule. It is commonly stated that in the earthworm carmine introduced into the gut is removed by the yellow cells, which then go free and pass out with their burden by the nephridia. In the Capitellidæ, and in *Tubifex*, on the other hand, the pigment is not wholly eliminated, but is in part stored in the skin. These facts may show that in worms the products of the metabolism of "liver" cells are not readily eliminated by the gut itself, and so may in some cases be important in coloration. It seems possible that in *Bonellia* also the colour of the skin is due to a modification of a pigment occurring in the gut (?), or in the cells lining the body cavity. Further, we have seen (p. 191) that in Mollusca the peculiar pigment enterochlorophyll, at least in some cases, colours the fæces ; that it occurs both in the cells of the gut and in the digestive gland ; and finally, that it is possible that in some cases it may give rise to the pigments colouring the mantle, and ultimately to those of the shell. Unfortunately this is as yet uncertain.

When we come to the fate of pigments introduced into the body cavity, and their natural analogues, the matter is much more difficult and complicated. In the following table the natural analogues column must be almost left blank :—

PIGMENTS INTRODUCED INTO BODY CAVITY

Organisms.	Situations in which introduced Pigments are found.	Natural Analogues.	Observer.
Polyzoa.	In the interior of leucocytes which remain within the zoœcium.		Harmer.
Echinoderms.	In wandering cells which may leave the body at any point, and which give rise to temporary pigmentation.	Similar wandering cells which contain pigment, and may give rise to pigmentation.	Durham.
Dytiscus (Insecta).	In amœboid blood corpuscles which later formed patches at various points in the skin.		Durham.
Crustacea.	In excretory organs.		Cuénot.

Conclusions

Although certain analogies between the fate of introduced pigments and the natural occurrence of pigment in the tissues are thus rare, yet suggestions as to analogy have been very freely made. Eisig considers that the pigment which occurs somewhat sparingly in the skin of the Capitellidæ is an effete product temporarily stored there, and further regards this as a widespread origin of pigment. Mr. Durham, basing his view largely on his own researches on Echinoderms, regards colouring - matters as either waste products or effete respiratory pigments which, when eliminated by means of amœboid cells, may give rise to coloration of the skin.

Of all suggestions of this kind, those of List for Vertebrates are the clearest and most definite, and may be summarised here.

In the first place, List accepts without reservation the view that pigment does not originate in epidermal structures, but is carried to them by wandering leucocytes. He believes that the pigment originates within the blood-vessels by the degeneration of red blood corpuscles ; that it is taken up by leucocytes ; and that these with their burden follow the track of the blood-vessels outwards from the corium to the sub-epithelial layer. The pigment granules are to be looked upon as excretory products, which are in part taken up by the epithelial cells and gradually eliminated as these degenerate. In bony fish, and apparently in Amphibia, pigment arises in the embryos from the degeneration of the yolk, but the pigment which is produced later probably arises from blood pigment.

Similarly the migration outwards of pigmented cells in the leech is often regarded as a process of excretion.

CRITICISM OF THESE CONCLUSIONS

As to the whole question, it is probably too soon to attempt to draw conclusions, but one or two points may at least be touched on. In the first place, an obvious difficulty in the way of regarding all pigments as waste products, or as derivatives of respiratory pigments, is that the great majority of the researches hitherto carried on have almost entirely omitted to consider the pigments soluble in alcohol.

Most have been conducted by the method of sections, and during the course of preparation of the objects, the lipochromes, and the numerous other unstable or soluble pigments, are completely removed or destroyed, so that of these the investigators have nothing to say. It is obvious, however, that it is these brightly coloured substances which give rise to the most striking of the phenomena of coloration. Practically any substance occurring in opaque granules may give rise to dull brownish colours, and so may be termed a pigment, but does this help us as to the origin of the bright blue of many jelly-fish, the gorgeous red of some Crustacea, the bright colours of fishes and birds? These may be " waste products," but there is yet no proof of it; they may be reserves; they may be comparable to the production of aniline dyes in the coal-gas industry, *i.e.* by-products (Durham), but there is as yet little certainty. It is possible that some of the difficulties may be solved by a careful study of the chætopterin group of pigments, for the members of it are widely distributed, tend to occur in connection with endodermic (digestive) organs, and under artificial conditions give rise to brightly coloured derivatives, but the investigations have still to be made.

Again, the method of study by means of injections has obviously its limitations as a method of determining the physiological value of pigments. Thus Cuénot found that introduced pigments in the case of the Crustacea were eliminated by the excretory organs, or by the hepatic cells and the fæces; they were never stored up in the epidermal tissues, and yet the Crustacea are remarkable for the profuse pigmentation of the epidermis and cuticle, and there is certainly a marked elimination of pigment in the shell at the moult. The elimination of introduced pigment by the skin in the Capitellidæ is regarded as evidence that the pigment naturally occurring there is a waste product, but the application of the same principle to the Crustacea is fraught with obvious difficulties. If conclusions are to be drawn from the fate of introduced pigments, then the pigment of the cuticle in Crustacea is not a waste product ; if, on the other hand, it is the elimination of pigment by a moult which is the criterion, then the pigment is a waste product.

As a whole, therefore, it would seem that while it is impossible for a physiologist to conceive of pigment being produced in the organism in the haphazard fashion some would have us believe, yet it is at present also impossible to give a universal physiological explanation of its origin ; it probably arises in many different ways. As yet the classification of pigments given in the second chapter cannot apparently be simplified.

CHAPTER XV

THE RELATION OF FACTS TO THEORIES

General Summary—Theories as to Origin of Colour : Poulton, Wallace, Eimer, Cunningham, Simroth — Criticism of Natural Selection—Criticism of Other Theories—Conclusion.

WE have now completed our general survey of the colours and colouring-matters of organisms. We have seen that these colours are due either to definite pigments deposited in the tissues, or to optical effects produced by the structure of these tissues. We have discussed the chemical, and, so far as is known, the physiological nature of some of the chief pigments, and described the appearances presented by the most striking optical colours. Finally, we have rapidly surveyed the colour phenomena presented by the most familiar plants and animals. That the survey as a whole is halting and incomplete must be obvious to all. We have seen that it is as yet impossible to give a definite physiological explanation of the origin of pigment; that it is practically impossible to classify pigments in a logical manner; that most of the problems connected with the subject are entirely unsolved. What is the meaning of the great series of lipochromes in the economy of animals? How do they arise, and why are they sometimes introduced, and in other cases synthetically formed? Why should they so frequently occur in pairs, and what is the relation between the red and the yellow series? These are only a few of the unanswered questions which make one at times doubtful whether it is not still too soon to attempt a synthetic survey of the biological aspects of colour. It is, however, noticeable that if the physiology of pigments and colour is still in an embryonic condition, yet the speculative side of the subject has attained rank and rapid growth. It is impossible to conclude a work of this description without some reference to theories, but we should pass to the consideration of these with a full consciousness of the blanks in knowledge.

THEORIES AS TO ORIGIN OF COLOUR IN ORGANISMS

1. *The Darwinian Theory.*—So far we have considered organisms as if they were isolated objects, uninfluenced by their surroundings; it is, however, one of the most striking characteristics of modern scientific thought that organisms are no longer looked upon as independent creations, but as linked to one another by the closest of relations. Their colours are often their most striking external features; we must ask what effect these colours have upon their relations to other organisms. Now it is a matter of common observation that the colours of some animals correspond so closely to the colours of the objects among

which they live, that they ca only be distinguished with difficulty. If the enemies of such species are psychologically similar to ourselves, the colouring must render them less conspicuous to these enemies, and must thus be protective. Therefore it may be said that, however the colour in these cases first arose, it must always have been, other things being equal, useful to the species ; therefore the forms displaying these colours would tend to persist, the others would tend to be eliminated ; therefore we may say that the colour arose by Natural Selection, which weeded out all those not possessing it.

This is in essence the explanation of colour phenomena given by a great number of naturalists at the present time. Colour they say is originally non-significant, a result of the chemical or physical properties of substances ; its appearance in the superficial tissues may render the organism better fitted to survive in the struggle for existence, and therefore is encouraged and maintained by Natural Selection. The various types of coloration presumed to be of use have been classified under the headings of Protective Resemblance, Mimicry, Warning Colours, and so on ; their use is supposed to be to protect the organism from its enemies, to enable it to steal unperceived on its prey, or to warn its enemies that it is unpalatable or dangerous and must be avoided. Beside these, however, there is another series of colours to be considered. We have already seen that in birds the males are frequently far more brilliantly coloured and ornamented than the females. As these colours do not fall into any of the divisions already mentioned, many naturalists have adopted

Darwin's view that they are due to the persistent choice by the females of the most ornamental males, and therefore to Sexual Selection. So that to the general statement that the colours of animals are due to the action of Natural Selection, we must add, except in the case of the bright colours of males, which are due to the action of Sexual Selection. This explanation of the colours of animals is substantially that given by Prof. Poulton in his *Colours of Animals*. Mr. Poulton is indeed one of the most thoroughgoing of all the adherents to the doctrine of Natural Selection, as the following example taken from his book may serve to show. He describes the buff-tip moth (*Pygœra bucephala*) as exhibiting a very marked resemblance to a broken piece of lichen-covered stick, and then come the following sentences :—" A friend has raised the objection that the moth resembles a piece of stick cut cleanly at both ends, an object which is never seen in nature. The reply is that the purple and gray colour of the sides of the moth, together with the pale yellow tint of the parts which suggest the broken ends, present a m t perfect resemblance to wood in which decay has induced that peculiar texture in which the tissu. breaks shortly and sharply, as if cut, on the application of slight pressure or the force of an insignificant blow " (*Colours of Animals*, p. 57). These statements, whatever else they do, certainly display a most profound faith in the efficiency of Natural Selection as a factor in evolution. The efficiency in this case seems almost excessive ; one cannot help wondering whether a protective resemblance which was a little less laboured would not have served the

purpose.   Another example of a similar elaboration
of protective resemblance may be also quoted from
Mr. Poulton's pages.   The insect in this case was
found by Mr. W. L. Sclater in Tropical America.   In
the place where it was found the leaf-cutting ants are
extremely numerous, and are constantly seen carrying
pieces of leaves "about the size of a sixpence held
vertically in the jaws."   The insect found by Mr.
Sclater, though not an ant, resembled one ; and,
moreover, had an anterior, thin, flat expansion which
imitated the leaf carried by the ants, so that, as a
whole, in Mr. Poulton's words, the insect " mimicked
the ant, *together with its leafy burden*" (*ibid.* pp.
252, 253).   Now, as it is only the homeward-bound
ants which carry pieces of leaves, it seems   in
this case also that the protective resemblance is
unnecessarily laborious ; something less might surely
have served.

Apart from this, however, the examples show how
some naturalists attack the problems of colour.   It
is unnecessary here to go into further detail as to the
various applications of the theory ; most of these have
now become completely popularised.

2. *Mr. Wallace's Theory.*—We shall next pass on
to consider the modification of this theory which is
supported by Mr. A. R. Wallace.   Mr. Wallace, in his
book on Darwinism, expresses his general belief in
the theory of colour production implicit in such terms
as Mimicry, Warning Coloration, etc., and dissents
only from the theory of the origin of the bright
colours of males by Sexual Selection.   In point of
fact, however, his dissent in reality carries him 'further
than this, and to some extent at least shakes the

whole theory of the origin of colour as a result of the action of Natural Selection.   Mr. Wallace, as is well known, gives up Sexual Selection on the ground that there is no evidence that the females do exercise such a selection ; while if they did, the effect of their choice would be neutralised by the action of Natural Selection.   The fact that the males in most animals are more brightly coloured than the females, Mr. Wallace ascribes in general terms to the " greater vigour and excitability of the male " ; if the colour and ornamentation be an expression of abundant vitality, its persistence and increase is easily accounted for apart from the choice of the female.   The hypothesis of Sexual Selection is therefore as needless as it is unproved.   Mr. Wallace then sums up his theory of the origin of colour in five theses, of which the following is a brief abstract :—Colour arises as a necessary result of the complex chemical constitution of animal tissues ; it becomes more conspicuous and intense as external tissues become more complicated in structure ; it is probable that colour development takes place according to definite laws of growth ; finally, " the colours thus produced, and subject to much individual variation, have been modified in innumerable ways for the benefit of each species."   It is in this way that Protective Coloration, Mimicry, etc. have been produced.   Again, in the higher forms the male as compared with the female exhibits brilliant colours due to his greater vigour, while his mate has been kept plain by Natural Selection.

Now all this is very different from the statements made by Mr. Poulton.   Both certainly begin by saying that colour is originally non-significant ; but

Mr. Wallace speaks of laws of growth as determining the progressive changes seen in the development of feather-markings; while Mr. Poulton tells us that although pigments tend to occur in animals, it is by no means certain that they would have appeared on the surface apart from Natural Selection, and that they tend to disappear from the surface directly they cease to be useful.

Thus, according to the school which is usually known as the Darwinian, colour, wherever seen, is due to the favouring influence of Natural Selection, and is in some way useful to the species. In the view of the popularisers of the subject, it therefore becomes the main object of the naturalist to invent as ingenious an explanation as possible of the way in which it is useful. If the naturalist's powers of invention fail, though this happens but rarely, then the colour is non-significant, or better still, the animal has recently changed its habitat, and is no longer perfectly adapted to its environment. The theory is, therefore, perfectly complete and coherent, and persons refusing to accept it are at once stigmatised as laboratory-made scientists, ignorant of nature, and unworthy of the name of naturalist.

Mr. Wallace's modified views, if less capable of a *reductio ad absurdum*, are apparently less completely logical. As noticed by Professor Geddes and Mr. Thomson, in their *Evolution of Sex*, the denial of Sexual Selection has a considerable bearing upon Natural Selection in general. To illustrate this, we may take an example from humming-birds. The genus *Eustephanus* includes the species *E. galeritus* and *E. fernandensis*, in both of which the sexes differ considerably from one another.   The sexual differences are especially well marked in the latter, in which the male is of a bright brown-red colour, a tint comparatively rare among the humming-birds. The female exhibits colours of a more usual type, but is remarkable in possessing a brilliant metallic crest, an ornament which is very rare among female humming-birds.   The male of the other species is very like the female of *E. fernandensis*, except that his crest is red instead of green, while his own female is very plain, and without a crest.   Now, if Mr. Wallace admits that the brilliancy of the male *E. fernandensis*, as compared with his mate, is due to his greater vigour and vitality, surely it is not unreasonable to conclude that the general greater brilliancy of this species, as compared with *E. galeritus*, is due to its greater vigour and vitality.   In other words, bearing in mind that the male of *E. galeritus* is hardly more brightly coloured than the female of *E. fernandensis*, may we not say that the two species bear to each other, as regards vitality, the same relation as the male and female of *E. fernandensis* bear to one another ?   If this be granted, then surely, other things being equal, the coloration of a species bears some relation to its vitality—that is, it is primarily determined by the physiological condition of the organism and not forced upon it by the stress of environmental conditions (see the *Evolution of Sex*).   Again, if this be so, much of the elaborate treatment of colour phenomena in the early part of Mr. Wallace's book seems needless.   If we may account for the colours of many birds as the incidental consequences of physiological conditions, then surely we need no elaborate discussion of the possible uses of colour, for we see that it arises apart from usefulness, and *ergo* may persist apart from usefulness. This is the view put forward in the *Evolution of Sex*, where the colours of organisms are regarded as expressions of the constitution of the individual.

3. *Mr. Cunningham and Professor Eimer's Theories.*—Although the theories as to the origin of colour, adopted on the one hand by Mr. Poulton, and on the other by Mr. Wallace, are widely accepted among biologists, dissentients are not wanting, and are probably on the increase. Among the older theories, that dependent upon the acceptance of Lamarck's factor of an inheritance of acquired characters, has been vigorously maintained by Mr. Cunningham in this country, and Professor Eimer and a numerous school abroad. Professor Eimer's theories of the origin of colours and markings involve especially the conception that in this, as in other respects, evolution is a progression along definite lines determined by laws of growth which are the accumulative result of environmental stimuli ; the emphasis is, however, so laid upon the laws of growth that the fact that these involve an inheritance of the effects of environmental influence is apt to be lost sight of. The difference between this and the preceding theories is best indicated by a concrete example. We may take the vexed question as to the reason for the absence of pigment in cave-inhabiting animals. According to Mr. Poulton, animals which live in darkness are pale, because pigment would not be visible in these situations and is consequently no longer of any use to them : it is, therefore, no longer

maintained by Natural Selection, and *therefore* it disappears,—the last *therefore* being one of the great points of dispute. Mr. Cunningham, on the other hand, considers that pigment is, or at least was primarily produced by the action of light on the skin, and that cave-inhabiting animals are pale-coloured because there is no light to stimulate the development of pigment. According to him, light and pigment are directly related; according to others, light is not the cause of pigmentation, it only puts in motion the machinery produced in the organism by Natural Selection. We have already seen by what beautiful experiments Mr. Cunningham has endeavoured to support and prove his position as to the relation between light and colour.

4. *Dr. Simroth's Theory.*—Mr. Cunningham's position may be taken as typical of that taken up by those who refer variation to the inherited and cumulative effect of environmental influences, but as an elaboration of the same principle we may take up a paper recently published by Dr. Heinrich Simroth. Dr. Simroth is well known, not only by his concrete researches, but as an ingenious and fertile theorist, and his present paper, though vague and mystical, has yet considerable interest, and to some extent may serve as a type of many of the most recent theories as to colour production. Dr. Simroth's theory is, however, remarkable in displaying an absolute indifference to the facts of chemistry, which even among biologists is relatively rare. As papers of this kind are exceedingly difficult to interpret, it may be well to state clearly that although the following is an attempt to give a purely objective

summary, yet it may quite well be that the subjective element is far from being absent.

So far, then, as I understand Dr. Simroth, he refers all pigments back to a prime substance which is closely united to primitive protoplasm, and which has evolved along with primitive protoplasm, producing the simple spectral colours in the order of the spectrum, beginning at the red end. That is, red pigments are simpler in composition than those of green or violet colour, and tend to appear earlier, and to be particularly prominent in simple organisms. We may thus speak of an evolution of pigments corresponding to an evolution of organisms, and the red or. yellow pigments correspond to the simpler organisms. These red and yellow pigments have a simple chemical composition and a small molecular weight, and as we pass upwards and find the colours of the pigments changing, so also we find the chemical composition growing more complex and the molecular weight increasing in amount.

As to the causation of this evolution of pigment, Dr. Simroth refers primarily to the effects of light and warmth, but makes the following detailed suggestion as to the determining cause of the actual direction of evolution.

In the first place, he suggests that at an early stage in the world's history the atmosphere was so saturated with watery vapour, that it at first only allowed the red rays of the sun's light to pass through, and then, as the vapour gradually cleared away, the other rays, in the direction from the red to the violet, were able to penetrate.

Secondly, he believes that protoplasm is so constituted, that it responds differently to the varying stimuli of the separate rays. Thus it responds to the rays of long wave-length by the formation of simple pigments, and to those of short wave-length by the formation of more complex pigments, so that there is a relation between the molecular weight of the pigments produced and the wave-lengths of the rays producing them. If we combine this statement with the previous one as to the relation existing between the colour of a pigment and its molecular weight, it would seem that red light produces red pigments of simple composition ; violet light, violet pigments of complex composition, and so on. Further, the previous suggestion as to the gradual appearance of the rays, accounts for the order in which the pigments appear.

Before proceeding further with Dr. Simroth's theory, we may note that so far it is in its details largely an adding together of the suggestions of others. Thus, the suggestion that the action of red light on organisms is to cause them to produce red pigment, is merely the suggestion as to the photographic sensibility of living beings which has already been made in various quarters. As every one knows, the essence of the process of photography lies in the fact that certain chemical substances are extremely sensitive to the action of light. When the photographer exposes a plate to light in his camera, the sensitive substance with which it is covered is rapidly decomposed by the action of the light, and dark-coloured substances are produced. So great is this sensitiveness that the brightest rays of the incident light correspond to the darkest parts of the " negative " produced, and all the developments of modern photography are rendered possible. Now, if it is possible to obtain inorganic substances which are so extraordinarily sensitive to light, it is surely not impossible that organic substances, in their ordinary position within the organism, may display a similar sensitiveness, and therefore that pigment production may be the result of exposure to light. Further, as every one knows, one of the great objects of recent photographers has been to discover a method of photographing in colours—that is, of finding substances which react in such a manner to different rays of light as to themselves build up compounds having the same colour as the incident light. According to Herr Otto Wiener, certain compounds of silver chloride will do this ; and he suggests that organic substances may possess the same property, and that thus " protective " coloration may be accounted for. A caterpillar may be like its environment, because its skin photographs that environment by means of the sensitive compounds of its own tissues. So far, therefore, Simroth's theory is largely based upon Wiener's suggestion, though he carries it much further.

Again, Simroth's suggestion as to a relation between the colour of a pigment and its chemical composition has been made on a smaller scale by Urech, whose researches on the pigments of butterflies we have already quoted. Urech, in commenting on the fact that in the butterflies of the genus *Vanessa* the wings are at first white and the colours then develop in the order of the spectrum (yellow, orange, red, brown, black), suggests that there is a relation between the molecular weight of these pigments and their respective colours, and that this gradual development of colour in the history of the individual corresponds to the evolution of colour in the history of the race.

We must now return to a more detailed consideration of Simroth's paper. He supports his central thesis as to the origin of all pigments showing simple spectral colours from a prime substance by three arguments, which are not, however, very sharply differentiated from one another.

His first argument is based upon the modifications of the retinal purple in Vertebrates. As is well known, the rods of the retina of most Vertebrates contain a purple pigment known as rhodopsin or " sehpurpur," which, when exposed to light, undergoes a series of changes—becoming red, orange, yellow, and finally colourless. These modifications Simroth, so far as the author understands him, regards as evidence that all pigments are genetically related, and that one can be derived from another. He also lays especial stress upon the fact that red pigment is usually associated with the eye-spots of simple organisms, and that such organisms seem never to possess dark-coloured pigments. This he regards as evidence that pigments belonging to the less refrangible end of the spectrum tend to appear first.

The second argument is based upon the modifications of the lipochromes of plants. Simroth regards chlorophyll as the result of the modification of a lipochrome, a view for which, as we have seen, there is practically no evidence. He also believes that lipochromes are of great importance both as reserve stuffs and as oxygen carriers in the process of assimilation. When the metabolism of the cell is active, oxygen is withdrawn from the yellow pigment, and it becomes converted into green chlorophyll. When metabolism diminishes, the green chlorophyll becomes oxidised and is converted into a lipochrome, and thus the colours of autumnal leaves, of fruits, and of flowers are produced. The fact that chlorophyll is commonly regarded as a nitrogenous compound, which the lipochromes are certainly not, is nowhere alluded to. Lipochromes, Simroth regards as pigments of relatively great simplicity, especially characteristic of plants as the simpler organisms. When they occur in animals, they are to be looked upon as evidences of a primitive condition, though they may be utilised for purposes of warning coloration, mimicry, etc., such colours, according to the author, being always of simple nature. If lipochromes are, however, evidences of a primitive condition, it is difficult to understand why they should be so frequent in birds ; but the author does not touch upon this.

The third kind of evidence upon which Simroth bases his thesis is the order of appearance of the colours which either belong to the right half of the spectrum, or are not pigmental colours at all. Such pigments are characterised by their chemical complexity, and are associated with complex tissues. Thus the greater intensity of animal life expresses itself in the nature of animal pigments ; the masses of simple colours, like red, yellow, and green, which are so common in plants, being rare in animals (but Crustacea ?).

This is in outline Simroth's theory as to the origin of colour. His paper contains also numerous other suggestions which we cannot well discuss here. In general, he appears to believe that pigments are all related, that their development follows a definite order, and that their origin is due to the properties of protoplasm and the inherited stimuli of external conditions, such as light and warmth. On these, which he calls " inorganic " factors, he is inclined to lay much stress, especially in such cases as the colours of shells, in which he says there can be no question of adaptation.

The theory is interesting in spite of its vagueness, and is included here because it is in many respects typical of recent theories. As neither it nor the other suggestions similarly based upon an inheritance of the effects of environmental stimuli, have the extensive following possessed by the Natural Selectionist theory, we shall return, before going further, to a detailed discussion of the latter, beginning with the familiar subject of mimicry.

CRITICISM OF NATURAL SELECTION

The existence of colour resemblances between widely separated organisms, and that explanation of it which is implicit in the term Mimicry, have recently become almost universally familiar. The term in its present use was first employed by Mr. Bates, and his suggestions were adopted by Darwin, Wallace, and others, and have since been widely accepted. Criticism has, however, never been wanting, and in a recent paper M. M. C. Piepers brings forward some new facts of great interest in this connection. Piepers opposes altogether the idea of the action of Natural Selection in the matter, and remarks that the idea that the phenomenon is maintained by the accruing practical profit to the organism is one "essentially English."

1. *Mimicry.*—As is well known, the doctrine of mimicry among butterflies involves primarily the hypothesis that birds are the great enemies of diurnal butterflies, that certain families of butterflies, notably the Heliconidæ, the Danaidæ, and the Acræidæ, are not attacked by birds, and that therefore wherever these butterflies occur they are mimicked by non-protected butterflies. Piepers attacks the prime proposition that birds are the great enemies of butterflies, and then discusses in detail some of the so-called examples of mimicry.

As to the first point, it is admitted on all hands that the night-flying Lepidoptera are constantly eaten by birds, but with regard to the diurnal forms the question is different. Observations as to the actual pursuit of butterflies by birds are exceedingly few, although Bates and Wallace speak of finding scattered wings in the forest. M. Piepers, during more than thirty years' observation in India and the Malay, saw one or two isolated cases only, and he quotes other observers (Pryer, Skertchly, Scudder) as being equally or more unfortunate. As a whole, he concludes that although some birds may occasionally eat diurnal butterflies, there is as yet no evidence of that habitual, unvarying persecution which the theory of mimicry demands—a conclusion which is somewhat surprising to the outsider.

The author then proceeds to discuss various cases, of which the first is an example rather of protective coloration than of mimicry proper. It is the case of the colour-change of the caterpillar in the Sphingides from green to brown just before pupation. The caterpillar feeds among green leaves and is then green, but it forms a chrysalis in earth, and the brown colour has been held to be of protective importance during the period when the caterpillar is in search of a retreat. Piepers, however, observes that the period which elapses between the cessation of feeding and the formation of the chrysalis is exceedingly short, in some cases not more than some minutes ; that the brown is constant, while the tint of the earth varies and is often quite different ; and that as the caterpillar is necessarily moving all the time, a protective colour can hardly be of much avail. Finally, the colour-change is exceedingly common in the larvæ of Lepidoptera at this stage, and is probably due to a discoloration resulting from the drying up of the skin preparatory to its being shed. It occurs also in *Sphinx* larvæ which are brown to begin with and not green, though here the change is so slight as to be little noticed.

The next case taken up is that difficult one of the occurrence of several different forms among the females of certain *Papiliones* of India and the Malay, which was discovered and discussed by Mr. Wallace (*Trans. Linn. Soc.* xxv. ; see also *Contributions to the Theory of Natural Selection*, London, 1871). The case is a somewhat difficult one, in part because the names used involve in themselves an interpretation of the facts. So far as it is possible for one who is not an entomologist to judge, the following is an impartial statement of the case :—The genus *Papilio* in India and the Malay shows a marked tendency to develop varieties (species) of definitely limited geographical distribution, so that, for example, a variety (species) occurring in India is replaced by a closely related variety (species) in the Malay. Further, in many cases there occurs a marked polymorphism in the females, which expresses itself in the fact that they may resemble the males or may display large spatulate appendices to their hind wings which are entirely absent from the males, as well as other peculiar characters. This, however, occurs as a specific character in both males and females of other series of *Papiliones*, the result being that some of the females of one series may resemble (mimic) both sexes of other series. The following table will perhaps make this plain, the brackets indicate that the forms are " geographical species" (replacing species), the habitat being indicated at the left :—

	SERIES A.		SERIES B.
Java	P. Memnon. Polymorphic.	One female " mimics "	P. Coon. (Sexes similar).
India and Sumatra	P. Androgeos. Polymorphic.	One female " mimics "	P. Doubledayi. (Sexes similar).

	SERIES C.		SERIES D.
India	P. Theseus. Polymorphic.	One female " mimics "	P. Hector. (Sexes similar).
Malay	P. Pammon. Polymorphic.	One female " mimics " Another ,, ,,	{ P. Diphilus. P. Antiphus. (Sexes similar).

Mr. Wallace explained the case as a typical one of

mimicry, displaying itself only in the females, and not in all of these. According to M. Piepers, it is to be explained on the ground that the polymorphic forms represent successive stages in the transition between one monomorphic species and another. He is of opinion that the Papilionidæ of the Malay are, for the most part, descended from ancestors with large spatulate appendices to the hind wings, but that many have lost or are losing these. In the two series *P. Memnon-Androgeos* and *P. Theseus-Pammon*, the species are undergoing this transition ; the males and some of the female forms display the new type of structure, while certain of the female forms display in part ancestral traits. The resemblance between these female forms respectively and the other two series of butterflies (*P. Coon-Doubledayi* and *P. Hector-Diphilus-Antiphus*) is not so great in the field as in the study, and is merely due to the fact that these two series display a more primitive type of structure and coloration, one nearer to that displayed by the hypothetical ancestor of the Eastern Papilionidæ.

The whole question is considerably complicated by the great variability of all the *Papiliones*, which makes it practically impossible to distinguish between species and varieties ; while, on the other hand, the nomenclature employed has a considerable bearing on the question of mimicry, at least when represented in tabular form. The mimicking females of *P. Memnon* and *P. Androgeos* are exceedingly alike, so much so that they were formerly classed together as one species. The chief difference is that the Javan form has yellow spots, while the Indian form has reddish. In the species *P. Coon* and *P. Double-*

dayi the same contrast of colour is observed—yellow in the Javan *P. Coon*, and red in the Indian *P. Doubledayi.* Wallace regards this as a proof of mimicry, the mimicking forms varying as the mimicked forms vary; Piepers regards it as a response to similar geographical conditions, and denies specific value to the forms *Coon* and *Double-dayi*, as also to *Memnon* and *Androgeos.* The case shows considerable resemblance to the one which he next considers, and which may be briefly noticed. Among the Satyridæ there are two closely related species, *Paraga Egeria* and *P. Megæra*, both common in Western Europe, of which the former frequents shady woods and the latter exposed places, especially the neighbourhood of walls heated by the sun. The colour of the first is a dull brown, with dull yellow spots, of the second a bright reddish-orange. When traced southwards, however, the tint of *P. Egeria* deepens and approaches more closely to that of *P. Megæra.* In Java there are two species of *Junonia*, *J. Erigone*, and *J. Asteriæ*, belonging to the Nymphalidæ and certainly not closely related to the above, which have respectively similar habits and coloration. Here, then, is one of those cases of duplex " mimicry," so dear to the hearts of many, spoilt only by the trifling circumstance that the two sets are separated by the distance of nearly half the globe !

This paper has been quoted at such length, not because it is the only detailed criticism of mimicry extant, but because of the care with which it is done, and the apparent strength of its criticisms. Very few of the instances of mimicry have been subjected to such authoritative criticism ; and the fact that those here submitted have not stood the ordeal furnishes a strong presumption that a large number of the cases contained in the literature of the subject are likewise valueless.   The term mimicry is applied indiscriminately to all cases of colour resemblance ; many of these can certainly not be so explained, therefore we are justified in saying that at the present time the explanation of the facts of colour resemblance implicit in the use of the term " mimicry " is insufficient, or to use Mr. Sedgwick's term, inadequate.   Mr. Sedgwick's observations in regard to the cell-theory seem indeed eminently applicable throughout to the theory of mimicry.   " A theory to be of any value must explain the whole body of facts with which it deals.   If it falls short of this, it must be held to be insufficient or inadequate ; and when, at the same time, it is so masterful as to compel men to look at nature through its eyes, and to twist stubborn and unconformable facts into accord with its dogmas, then it becomes an instrument of mischief, and deserves condemnation, if only of the mild kind implied by the term inadequate " (" Remarks on the Cell Theory," *Compte Rendu*, 3me, Cong. Zool. Leyde, 1896, p. 121).

2. *Protective Resemblance.*—Although mimicry is commonly said to be merely a special case of protective colour resemblance, it is in some respects more difficult to understand than the latter, and it is not perhaps necessary to suppose that the two stand or fall together.   It is, of course, to be clearly understood that the existence of resemblances between

organisms and their surroundings or between un-
related organisms is denied by no one; it is the
explanation involved in the use of the terms " protective " and " mimicry," which is doubtful.

It would lead us too far to enter in detail into
all the arguments which have been advanced as
tending to prove that the resemblance between
organisms and their surroundings has been acquired
and is maintained by the aid of Natural Selection ;
the following summary of the facts in the case of
the Lepidoptera, taken from Weismann, is sufficient
for our purpose here.

Weismann states the case as follows :—" Im-
mune " butterflies, such as the Heliconidæ, the
Danaidæ, the Acræidæ, the Euplocidæ, have usually
both surfaces of their wings coloured alike, and never
resemble their surroundings in the resting position ;
unprotected butterflies, such as the Nymphalidæ, are
in the great majority of cases protectively coloured
on their lower surface. Further, the coloration of
this lower surface bears a close relation to the
position of the wings in repose—that is, if in this
position the hind wings overlap the fore, it is only
the exposed tip of the fore-wing which is protectively
coloured ; while if, as in *Kallima*; there is no overlapping, the whole under surface of the fore-wings
displays this type of coloration. Again, the special
type of protective coloration which consists in
resembling a leaf, is exceedingly common among
wood-inhabiting butterflies whether related or not.
This coloration bears no definite relation to the
structure of the wing, but " die Fläche behandelt als
eine *tabula rasa*, auf der man zeichnen kann, was

man will," the " man " in question being apparently the all-compelling force of Natural Selection.

The facts so stated are certainly sufficiently remarkable and seem at first sight at least to warrant Weismann's conclusion that they are only explicable by the action of Natural Selection, but detailed reflection shows many difficulties. The prime assertion of the immunity of the Danaidæ, etc. is denied by many, the persecution of the protectively coloured butterflies is also, as we have seen, doubted by field entomologists. The relation between the colour of the wings and the position taken up by them in repose seems a very striking fact, but I have noticed a somewhat similar occurrence in the feathers of birds. In the humming-bird *Cynolesbia gorgo* the tail is forked and the tail-feathers overlap one another ; the tips of the feathers are of a gorgeous metallic colour, but this is confined to a simple band at the exposed part, the part of the feather which is overlapped being a deep black. Owing to the forking of the tail, the overlapping is such that in each quill more of the vane is covered on one side than on the other, the distribution of black and metallic colour corresponds exactly to this overlapping, so that the metallic colour extends farther back on one side of the rachis than on the other. Here is a case very like that of the butterflies' wings, and yet it is almost impossible to believe that it can have been produced and maintained by its utility (see also p. 167).

We have confined our study of protective coloration and mimicry to the Lepidoptera, because they are admitted on all hands to exhibit the phenomena most strikingly. The two papers which have been chosen to represent the two positions in regard to the matter illustrate at least the main fact that both parties are somewhat stronger in attack than in defence. It would be easy to multiply references almost indefinitely, but this would in large part involve mere repetition. The advocate of the *Allmacht* of Natural Selection reïterates in many tones the well-established facts of colour resemblance, and the insufficiency of laws of growth, of correlation of parts, and the rest to account for these. His opponent returns to the charge again and again, well armed with the lack of evidence, the absence of experimental verification, the disproof of particular cases ; there are weak places in the walls of both citadels, but both parties are strong in attack ; all the clamour has not, however, as yet caused the walls of either Jericho to fall.

To drop the metaphor, it must be obvious from the above discussion that there are great difficulties in the acceptance of Natural Selection as the most important factor in the evolution of colour, and that there is little doubt that its aid has been invoked in far too reckless a fashion. At the same time, it must be confessed that there is not as yet in the field a complete and cogent theory which is capable of dispensing with Natural Selection ; whether this is due to ignorance of physiology, or to the real importance of this factor must be left to the future to decide.

CRITICISM OF OTHER THEORIES

Theories which attempt to minimise Natural Selection seem always sooner or later to assume an inheritance of acquired characters, and of this there is little evidence.   They also assume that environmental influences have a direct effect upon the organism, and of this Weismann's work has made many doubtful.   Or rather, Weismann has endeavoured to prove that those apparently direct responses to environmental stimuli which are facts of experience, can be interpreted also as the result of adaptation, and this, if proved, is fatal to theories like that of Simroth.   It is, however, to be noticed that in the case of the artificially produced variations in the colours of butterflies, competent entomologists (*e.g.* Garbowski) are of opinion that the new colours have little or no phylogenetic importance, and that as yet it is impossible to correctly interpret them.   There is, indeed, much evidence to show that in the case of butterflies the colours can be influenced by their surroundings in a way of which the mechanism is at present unknown.

Much of this is, however, apart from our main object, which is merely to show that in spite of the fluency with which so many people talk of the meaning of colour in organisms, the subject is as incomplete on the theoretical as on the physiological side.   It seems reasonable to believe that the two deficiencies are related, and that a little more physiology will arm the theorists with better weapons.   In the meantime, we cannot end a book on Colour more fitly than by an appeal for more facts.

REFERENCES

1. *General Works of Reference.*—For general facts and theories the following among others may be consulted :—

BEDDARD, F. E.—*Animal Coloration* (London, 1892).

EIMER, G. H. T.—*Organic Evolution, as the Result of the Inheritance of Acquired Characters, according to the Laws of Organic Growth.* Trans. by J. T. Cunningham (London, 1890).

GEDDES, P., and THOMSON, J. A.—*The Evolution of Sex* (Contemporary Science Series, London, 1889).

POULTON, E. B.—*The Colours of Animals* (International Science Series, London, 1890).

WALLACE, A. R.—*Darwinism* (London, 1889).

Reference should also, of course, be made to the works of Darwin, especially *The Descent of Man, and Selection in Relation to Sex* (London, 1871), and to the numerous books of travel which will be found cited in the above.

2. *Special Questions.*

ABEL, J., and DAVIS, W.—"The Pigments of the Negro's Skin and Hair," *Jour. Exper. Med.* vol. i. No. 3 (1896), pp. 361-400 1 pl.

AGASSIZ, A.—*The Cruise of the Blake*, Bull. Museum Compar. Zool. Harvard College, U.S.A., 1888. Many Observations on Colours of Marine Organisms.

ALCOCK, H.—"The Asteroidea of the Indian Marine Survey," *Ann. and Mag. Nat. Hist.* vol. xi. (1893), pp. 73-121, 2 pls. Colours of Deep-sea Forms.

ANDRÉ, E.—"Le Pigment mélanique des Limnées," *Revue Suisse de Zool.* iii. (1895), pp. 429-431.

BATESON, W.—*Materials for the Study of Variation*, London, 1894. Especially for Colour Variation, and for Colours of Pleuronectidæ.

BECQUEREL, HENRI, et BRONGNIART, CHARLES—"La matière verte chez les Phyllies," *C. R. Ac. Sci.* cxviii. No. 24 (1894), pp. 1299-1303.

BEDRIAGA, J. VON—" Mittheilungen über die Larven der Molche," *Zool. Anzeig.* xiv. (1891), pp. 295-300, 301-308, 317-323, 333-341, 349-355, 373-378, 397-404; and xviii. (1895), pp. 153-157. Development of Colour.

BOGANDOW, ANDRÉ—" Études sur les causes de la coloration des oiseaux," *Compt. Rend.* xlvi. (1858), pp. 780, 781.

BOYCE, R., and HERDMAN, W. A.—" On a Green Leucocytosis in Oysters associated with the Presence of Copper in the Leucocytes," *Proc. Roy. Soc.* lxii. (1897), pp. 30-38.

BRÜCKE—(Colours of Octopus), *Sitzungsb. d. Akad. Wiss.* Wien. viii. (1852), p. 196. Structural Colour.

BÜRGER, O.—*Die Nemertinen des Golfes von Neapel*, 22 Monographie, Berlin, 4to, xvi + 743 pp. 31 pls. For Colours.

BÜTSCHLI, O. — " Protozoa," Bronn's *Klassen u. Ord. d. Thierreichs*, Leipzig, 1889. Pigments of Different Forms.

CARAZZI, D.—"Contributo all' istologia e alla fisiologia del Lamellibranchi 1. Ricerche sulle ostriche verdi," *Mittheil. Stat. Neapel*, xii. (1896), pp. 381-431, 1 pl. See also letter in *Nature*, vol. lii. pp. 643, 644.

CHATIN, A., and MUNTZ, A.—" Étude chimique sur la nature et les causes du verdissement des Huîtres," *C. R. Ac. Sci.* vol. cxviii. (1894), pp. 17-23 and 56-58.

CHURCH, A. H.—" Researches on Turacin, an Animal Pigment containing Copper," *Chem. News*, vol. xix. (1869), p. 265. " Researches on Turacin, an Animal Pigment containing Copper," *Phil. Trans. Roy. Soc.* vol. clix. (1870), pp. 627-636. " Lecture on Turacin," reported in *Nature*, vol. xlviii. (1893). " Researches on Turacin," *Phil. Trans. Roy. Soc.* 183a, pp. 511-530.

COSTE, F. H. PERRY—" Insect Colours," *Nature*, vol. xlv. pp. 513-517, 541, 542, and 605, 1892.

CUÉNOT, L.—"Études physiologiques sur les crustacés Decapodes," *Arch. Biol.* xiii. (1894), pp. 245-303. Elimination of Introduced Pigments. " Études physiologiques sur les Orthoptères," *Arch. Biol.* xiv. (1895), pp. 293-341, 2 pls. Functions of Fatty Body, and Nature of Pigments.

(a) CUNNINGHAM, J. T.—" Researches on the Coloration of Flatfishes," *J. Mar. Biol. Ass.* iii. (1893), pp. 111-118.

(b) CUNNINGHAM, J. T., and MACMUNN, C. A.—" On the Coloration of the Skins of Fishes, especially of Pleuronectidæ," *Phil. Trans. Roy. Soc.* London, vol. clxxxiv., B. pp. 765-812, 3 pls., 1893.

DELÉPINE, SHERIDAN—" Proceedings of Physiological Soc.," *Journ. Physiol.* xi. (1890), p. 27. Origin of Melanin.

DURHAM, HERBERT E.—" On Wandering Cells in Echinoderms," *Q. J. M. S.* vol. xxxiii. (1891), pp. 81-121, 1 pl. Origin of Pigment.

EHRING, C. — *Ueber den Farbstoff der Tomate (Lycopersicum esculentum), Ein Beitrag zur Kenntnis des Carotins*, Münster, 1896, pp. 1-35.

EIMER, THEODOR—*Ueber das Variiren der Mauereidechse*, Berlin, 1881. *Ueber die Zeichnung der Thiere*, Humboldt, 1885, 1886, 1887, 1888.

EISIG, H.—*Die Capitelliden*, Naples Monograph, 1887. Pigment and Waste Products.

EMERY, C.—"Untersuchungen über *Luciola italica*," *Zeitschr. wiss. Zool.* xl. (1884), pp. 338-354, 1 pl. (Luminosity of *Luciola italica*), *Bull. Soc. Entomol. Ital.* xviii. (1885), pp. 351-355, 1 pl.

ENGELMANN, W.—"Farbstoff von *Hæmatococcus*," *Bot. Zeit.* xxxix. 1882. "Farbe und Assimilation," *Botan. Zeit.* 1883, pp. 1-17, and 17-29. "Die Purpurbakterien und ihre Beziehung zum Licht," *Bot. Zeit.* 1888, pp. 42-46.

FAXON, W.—"The Stalk-eyed Crustacea," *Mem. Mus. Harvard*, xviii. (1896), pp. 1-292, 67 pls. Chapter on Colour.

FISCHER, E.—*Transmutation der Schmetterlinge infolge Temperaturänderungen. Experimentelle Untersuchungen uber die Phylogenese der Vanessen*, Berlin, 8vo, 36 pp., 1895.

FLEMING, W.—"Ueber den Einfluss des Lichts auf die Pigmentirung der Salamander-larve," *Arch. Mikr. Anat.* xlviii. (1896), pp. 369-374.

FRANCIS, G.—"Pigments in Fishes' Skins," *Nature*, vol. xiii. p. 167.

FRITSCH, ANTON—"Über Schmuckfarben bei *Holopedium gibberum*," *Zoolog. Anzeig.* xiv. (1891), pp. 152, 153.

GADOW, H.—"The Coloration of Feathers as affected by Structure," *Proc. Zool. Soc.* 1882, pp. 409-421, 2 pls. See also Bronn's *Thierreich*, vi. 4, pp. 575-584.

GARBOWSKI, T.—"Descendenztheoretisches über Lepidopteren," *Biol. Centralbl.* xv. (1895), pp. 657-672. Summary and Criticism of Experiments of Fischer and Others.

GÄTKE, H.—*Heligoland as an Ornithological Observatory: the Result of Fifty Years' Experience.* Trans. by R. Rosenstock, Edinburgh and London, 8vo, pp. x. + 599, 1895. For Change of Colour without Moult.

GEDDES, PATRICK—"Further Researches on Animals containing Chlorophyll," *Nature*, 1882, pp. 303-305. "On the Nature and Functions of the Yellow Cells of Radiolarians," *Proc. Roy. Soc. Edinburgh*, vol. xi. (1881-1882), pp. 377-396.

GESSARD—"*Bacillus pyocyaneus*," *An. d. l'Institut Pasteur*, 1891, p. 65. Blue Colouring-matter.

GIESBRECHT, W.—"Mittheil. ü. Copepoden," *Mittheil. Stat. Neapel*, xi. (1895), pp. 631-694, 1 Fig. For Luminosity.

GIROD, P.—*C. R. Ac. Sci.* xciii. p. 96; and *Arch. Zool. exp. et gen.* ix. (1882), pp. 1-100. Analysis of Melanin.

GRAF, ARNOLD—"Über den Ursprung des Pigments ü. der Zeichnung bei den Hirudineen," *Zool. Anzeig.* xviii. (1895), pp. 65-70.

GREENWOOD, M.—"The Process of Digestion in *Hydra fusca*," *Journ. Physiol.* ix. Meaning of Brown Pigment.

HÄCKER, V.—"Untersuchungen über die Zeichnung der Vogelfedern," *Zool. Jahrbüch*, iii. (1888), pp. 309-316, 1 pl.

HALIBURTON, W. D.—"On the Blood of Decapod Crustacea," *Jour. Physiol.* vi. (1883), pp. 300-335, 1 pl. Pigment of Hypoderm of Lobster.

HARMER, SIDNEY J.—"On the Nature of the Excretory Processes in Marine Polyzoa," *Q. J. M. S.* vol. xxxiii. (1891), pp. 123-167, 2 pls. Formation of Brown Body.

HERDMAN, W. A.—"Note on a New British Echiuroid Gephyrean, with Remarks on the Genera *Thalassema* and *Hamingia*," *Q. J. M. S.* xl. (1898), pp. 367-384, 2 pls. *Re* Green Pigment.

HOPKINS, F. GOWLAND—"Pigment in Yellow Butterflies," *Abst. Proc. Chem. Soc.* vol. v. p. 117 (1889). See also *Nature*, vol. xl. p. 335, and letter in *Nature*, vol. xlv. pp. 197, 198 (1891). "Pigments of Lepidoptera," *Nature*, vol. xlv. p. 581 (1892). "The Pigments of the Pieridæ," *Proc. Roy. Soc.* London, vol. lvii. pp. 5, 6 (1894); and *Phil. Trans.* clxxxvi. (1896), pp. 661-682.

JOHNSON, H. P.—"A Contribution to the Morphology and Biology of the Stentors," *Journ. Morphol.* viii. (1893), pp. 468-552, 4 pls. Pigment of *S. cærulea*.

JONES, E. LLOYD—"Observations on the Specific Gravity of the Blood," *Journ. Physiol.* vol. xii. (1891), pp. 299-346, 4 pls. Relation to Colour of Skin and Iris.

KARAWAIEW, W.—"Über *Aulacantha scolymantha*," *Zoolog. Anzeig.* xviii. (1895), pp. 293-301. Function of Phæodia.

KEEBLE, F. W.—"The Red Pigment of Flowering Plants," *Sci. Progress*, New Series, vol. i. (1897), pp. 406-423. Abstract of Observations of Stahl and Others.

KEELER, CHARLES—"Evolution of the Colours of North American Land-birds," *P. Calif. Ac.* iii. (1893), xii + 361 pp. 19 pls.

KERSCHNER, LUDWIG—"Zur Zeichnung der Vogelfeder," *Zeitschr. Wiss. Zool.* (1886).

KNAUTHE, KARL—"Erfahrungen über das Verhalten von Amphibien u. Fischen gegenüber der Kälte," *Zool. Anzeig.* xiv. (1891), pp. 109-115. Change of Colour in Cold.

KÖLLIKER, A.—"Ueber die Entstehung des Pigments in den Oberhautgebilden," *Zeit. Wiss. Zool.* xlv. (1887), pp. 713-717, 2 pls.

KRUKENBERG, C. FR. W.—*Vergleichend-physiologischen Studien*, Heidelberg, 1880-1882. Contain numerous Papers on Animal Pigments. *Grundzüge einer vergleichenden Physiologie der Farbstoffe u. der Farben*, Heidelberg, 1884, pp. 85-184. With copious Bibliography.

KÜKENTHAL, WILLY—"Ergebnisse einer Zoologischer Forschungsreise in den Molukken u. in Borneo," *Abhand. Senckenberg. Nat. Ges. Frankfurt a. M.* xxii. (1896). Re *Ueber die Färbung der Tiere unter spezieller Berücksichtigung der tropischen Formen*, pp. 53-62.

LANGERHANS, PAUL—"Die Wurmfauna von Madeira," iv., *Zeit. Wiss. Zool.* vol. xl. (1884), pp. 247-285, 3 pls. Colours of Various Forms.

LANKESTER, E. RAY—"Report on the Spectroscopic Examination of certain Animal Substances," *Jour. Anat. and Physiol.* vol. iv. (1870), pp. 119-129. "A Contribution to the Knowledge of Hæmoglobin," *Proc. Roy. Soc.* vol. xxi. pp. 70-81, 1 pl. (1873). "Green Oysters," *Quart. Journ. Micr. Sci.* vol. xxvi. pp. 71-93, 1 pl. (1886). "Blue Stentorin, the Colouring-matter of *Stentor cæruleus*," *Q. J. M. S.* xiii. (1873), pp. 139-142. "On a Peach-coloured Bacterium," *Q. J. M. S.* xiii. (1873), pp. 408-425, 2 pls. "On the Green Pigment of the Intestinal Wall of the Annelid Chætopterus," *Q. J. M. S.* xl. (1897), pp. 447-468, 4 pls.

LEYDIG—"Die Pigmente der Hautdecke u. der Iris," *Verhandl. d. Phys. Med. Gesell.* Würzburg, xxii. (1888).

LINDEN, MARIA V.—"Die Entwicklung der Skulptur u. der Zeichnung bei den Gehäuse Schnecken des Meeres," *Zeitschr. f. Wiss. Zool.* lxi. (1896), pp. 261-317, 1 pl.

LIST, H. J.—"Ueber die Herkunft des Pigmentes in der Oberhaut," *Biolog. Centralbl.* x. (1891), pp. 22-32.

LOEB, J.—"A Contribution to the Physiology of Coloration in Animals," *Journ. Morphol.* viii. (1893), pp. 161-164.

MACALLUM, A. B.—"On the Distribution of Assimilated Iron Compounds, other than Hæmoglobin and Hæmatins, in Animal and Vegetable Cells," *Q. J. M. S.* xxxviii. (1895), pp. 175-274, 2 pls., *P. R. Soc.* London, lvii. pp. 261, 262.

MACINTOSH, W. C.—"A Monograph of British Annelida," *Roy. Soc.* vol. xxxiv. For Colours of Nemertines.

M'KENDRICK, J.—"On the Colouring-matter of Medusæ," *Journ. of Anat. and Physiol.* vol. xv. (1881), pp. 261-264.

MACMUNN, C. A.—"Studies in Animal Chromatology," *Proc. Birm. Philos. Soc.* vol. iii. (1883). "Observations on the Colouring-matters of the so-called Bile of Invertebrates," *Proc. Roy. Soc.* London, xxxv. (1883), pp. 370-403, 1 pl. "The Chromatology of the Blood of some Invertebrates," *Quart. Journ. Micr. Sci.* vol. xxv. pp. 469-489, 2 pls. (1885). See also other papers referred to in this article. "Observations on the Chromatology of the Actiniæ," *Trans. Roy. Soc.* London, clxxvi. (1885), pp. 641-663, 2 pls. "Further Observations on Enterochlorophyll and Allied Pigments," *Trans. Roy. Soc.* London, clxxvii. (1) 1886, pp. 235-266, 2 pls. "Contributions to Animal Chromatology," *Q. J. M. S.* xxx. (1889), pp. 51-96, 1 pl.

MALARD, A. E.—"The Influence of Light on the Coloration of the Crustacea," *Bull. de la Soc. Phil. de Paris*, 8, iv. (1892), pp. 24-30. Translated in *Ann. and Mag. Nat. Hist.* vol. xi. pp. 142-149.

MALY, R.—"Ueber die Dotterpigmente," *Akad. der Wiss.* Wien. lxxxiii. (1881).

MAYER, A. G.—"The Development of the Wing Scales and their Pigments in Butterflies and Moths," *Bull. Mus. Comp. Zool. Harvard*, vol. xxix. (1896), pp. 209-236, 7 pls. "On the Colour and Colour-Patterns of Moths and Butterflies," *Proc. Boston Soc. Nat. Hist.* vol.

xxvii. (1897), pp. 243-330, 10 pls.　See also *Bull. Mus. Comp. Zool. Harvard*, vol. xxx. (1897), pp. 169-256, 10 pls.

MAYER, P.—"Ueber das Farben mit Carmin, Cochenille u. Hämatein Thonerde," *Mit. Zool. Stat. Neapel.* x. (1892), pp. 480-504 ; and "Zur Kenntnis von *Coccus cacti*," *Ibid.* pp. 505-518, 1 pl.

MEYER, A. B.—"Ueber einen bemerkenswerthen Farbenunterschied der Geschlechter bei der Papageien-Gattung *Eclectus*," *Verh. d. k. k. Zool-bot. Ges.* Wien, 1874.

MOLISCH, H.—"Ueber. den Farbenwechsel anthokyanhältiger Blätter bei rasch eintretendem Tode," *Bot. Zeit.* xlvii. (1889), pp. 17-23. "Das Phycocyan, ein krystallisibar Eiweisskorper," *Botan. Zeit.* vol. liii. (1895), p. 131.

MOSELEY, H. N.—"On the Colouring-matter of Various Animals," *Q. J. M. S.* xvii. (1877), pp. 1-23, 2 pls.

NATHUSIUS, W. VON—"Ueber Farben der Vogeleier," *Zool. Anzeig.* xvii. (1894), pp. 440-445, and 449-452, 2 Figs.

NENCKI, M.—"Ueber die biologischen Beziehungen des Blatt u. Blutfarbstoffs," *Ber. de Chem. Gesell.* Bd. xxix. p. 2877.

NEWBIGIN, M. I.—"The Pigments of Animals," *Nat. Sci.* viii. (1896), pp. 94-100 and 173-177. "Observations on the Metallic Colours of the Trochilidæ and the Nectariniidæ," *Proc. Zool. Soc.* (1896), pp. 283-296, 2 pls. "An Attempt to Classify Common Plant Pigments, with some Observations on the Meaning of Colour in Plants," *Trans. and Proc. Bot. Soc.* Edinburgh, lix. (1896), pp. 534-550. "The Pigments of the Decapod Crustacea," *Jour. Physiol.* xxi. (1897), pp. 237-257. "The Pigments of the Salmon," *Report on Salmon, Fishery Board for Scotland* (1898), pp. 159-164. "On Certain Green (Chlorophylloid) Pigments in Invertebrates," *Q. J. M. S.* xli. (1898).

NÜSSLIN, O.—"Ueber einige neue Urthiere aus dem Herrenwieser See im badischen Schwarzwalde," *Zeit. wiss. Zool.* xl. (1884), pp. 697-724, 2 pls.

PANCERI, PAUL—"On the Luminous Organs and Light of *Pyrosoma*," *Q. J. M. S.* vol. xiii. (1873), pp. 45-51. "On the Light of *Phyllirrhoë bucephala*," *Ibid.* pp. 109-116.

PHISALIX, C.—"Recherches sur la matière pigmentaire rouge de *Pyrrhocoris apterus*," *C. R. Ac. Sci.* cxviii. pp. 1282-83.

PIEPERS, M. C.—"Mimétisme," *C. R. Third Zool. Congress*, Leyden (1896), pp. 460-476, 2 tables.

POULTON, E. B.—"The Essential Nature of the Colouring of Phytophagous Larvæ (and their Pupæ) ; with an Account of some Experiments upon the Relations between the Colour of such Larvæ and that of their Food-plants," *Proc. Roy. Soc.* London, vol. xxxviii. pp. 269-314, 1 Fig. (1885). "The Experimental Proof that the Colours of Certain Lepidopterous Larvæ are largely due to Modified Plant Pigments derived from Food," *Proc. Roy. Soc.* London, vol. liv. pp. 417-430, 2 pls. (1893).

ROSTAFINSKI — " Ueber den rothen Farbstoff einiger Chlorophyceen, etc.," *Bot. Zeit.* (1881), p. 463.

SCHAUDINN, FRITZ—" *Myxotheca arenilega,* nov. gen. et spec.," *Zeitschr. wiss. Zool.* lvii. (1893), pp. 17-31, 1 pl. Describes brilliant Red Pigment.

SCHENKLING, C.—"Mutmasslicher Farbenwechsel der Vogelfeder ohne Mauser," *Biolog. Centralbl.* xvii. (1897), pp. 65-79.

SCHUBERG, A.—"Zur Kenntniss des *Stentor cæruleus,*" *Zool. Jahrb.* iv. (1890), pp. 197-238. For Blue Pigment.

SCUDDER—*The Butterflies of the Eastern United States and Canada with Special Reference to New England,* Cambridge, three vols. pp. 24 + 1958, 89 pls. Colour and Markings.

SEMON, R.—" Entstehung u. Bedeutung der embryonalen Hüllen u. Anhangsorgane der Wirbelthiere," *C. R. Third Zool. Congress,* Leyden (1895). For Relation between Progress and Elimination of Nitrogenous Waste.

SIMROTH, HEINRICH—"Ueber die einfachen Farben im Thierreich," *Biolog. Centralbl.* xvi. (1896), pp. 33-51.

SLATER, C.—"On a Pigment-forming Bacterium," *Q. J. M. S.* xxxii. (1891), pp. 409-416, 1 pl.

SORBY, H. C.—"On the Colouring-matter of *Bonellia viridis,*" *Q. J. M. S.* xv. (1875), pp. 166-172. "The Colours of the Eggs of Birds," *Proc. Zool. Soc.* (1875), pp. 351-365.

SPULER, A.—" Beiträge zur Kenntnis des feinen Baues und der Phylogenie der Flügelbedeckung der Schmetterlinge," *Zool. Jahrbuch. Abth. Anat.* vol. viii. pp. 520-543, 1 pl. (1895). Includes numerous references.

STAHL, E.—"Über bunte Laubblatter," *Ext. Ann. Jardin Botanique de Buitenzorg,* vol. xiii. pp. 137-216, 2 pls. Abstract in *Bot. Zeit.* xiv. (1896), pp. 209-215.

SWINHOE, C.—"On the Mimetic Forms of Certain Butterflies of the Genus *Hypolimnas*" (Abstract), *P. R. Soc.* London, liii. (1893), p. 47.

TASCHENBERG, O.—"Einige Bemerkungen gegen H. Wickmann," *Zool. Anzeig.* xvii. (1894), pp. 304-309. Colours of Birds' Eggs.

URECH, F.—"Chemisch-analytische Untersuchungen an lebenden Raupen, Puppen u. Schmetterlingen u. an ihren Secreten," *Zool. Anzeig.* vol. xiii. pp. 254-260, 272-280, 309-314, and 334-341, (1890). "Beobachtungen über die verschiedenen Schuppenfarben u. die zeitliche Succession ihres Auftretens (Farbenfelderung) auf den Puppenflügelchen von *Vanessa urticæ* u. *io,*" *Zool. Anzeig.* vol. xiv. pp. 466-473 (1891). "Beiträge zur Kenntnis der Farbe von Insektenschuppen," *Zeitschr. wissenschaft. Zool.* vol. lvii. pp. 306-384, (1893). "Beobachtungen von Compensationsvorgängen in der Farbenzeichnung bezw. unter dem Schuppenfarben an durch thermische Einwirkungen enstandenen Aberrationen und Subspecies einiger Vanessa-Arten. Erwägungen darüber u. über die phyletische Recapitulation der Farbenfelderung in d. Ontogenese," *Zool. Anzeig.* xix. (1896), pp. 163-174 and 177-185.

VANHÖFFEN—"Das Leuchten von *Metridia longa*," *Zool. Anzeig.* xviii. (1895), pp. 304, 305.

WALTER, B.—*Die Oberflächen oder Schillerfarben*, Braunschweig, pp. vi. and 122, 8 Figs., 1 pl. (1895).

WEISMANN, AUGUST—"Ueber Germinal-Selektion," *C. R. Third Zool. Congress*, Leyden (1895), pp. 35-70. Protective Coloration in Butterflies.

WICKMANN, H.—*Die Färbung der Vogeleier*, Münster, 64 pp. (1893).

WIENER, OTTO—"Farbenphotographie durch Körperfarben u. mechanische Farbenanpassung in der Natur," *Ann. Phys. u. Chem.* lv. (1895), pp. 225-281.

ZENNECK, J.—"Die Anlage der Zeichnung u. deren physiologische Ursache bei Ringelnatter-embryonen," *Zeitschr. wiss. Zool.* lviii. (1894), pp. 364-393, 1 pl.

ZOPF, W.—*Beitrage zur Physiologie u. Morphologie niederer Organismen*, Leipzig, Heften i. ii. iii. (1892-93). "Ueber Pilzfarbstoffe," *Bot. Zeit.* xlvii. (1889), pp. 85-92. *Ibid. Ber. Deutsch. Bot. Gesell.* ix. (1891), pp. 22-28, 1 Fig.

INDEX OF AUTHORS

INDEX OF SUBJECTS

Printed by R. & R. CLARK, LIMITED, *Edinburgh.*

Lightning Source UK Ltd.
Milton Keynes UK
UKHW012249140219
337323UK00011B/667/P